ADVANCE PRAISE FOR

Resisting Nudities: A Study in the Aesthetics of Eroticism

"In her well-documented cultural and philosophical book, Florence Dee Boodakian warns us against the danger of a society of *auto-surveillance*, a coded society where the individual has gradually and naturally internalized rules that make freedom an illusion, depriving him/her of being a 'subject' in the unknown (and rich) land of eroticism. In addition, she guides us with great panache toward this no man's land of the unconscious that separates the erotic from pornography."

Isabelle Pagot-Votelet
Professor, IAV d'Orleans (Institute of Visual Arts)
and Sefco, University of Orleans, France

"Florence Dee Boodakian's *Resisting Nudities* proposes a new appreciation of eroticism based on a re-conception of what eroticism is. The erotic is that which interrupts itself, producing desire through moments of the 'discontinuous' that promise a continuity of ground between subject and object that is never there. The erotic is the fugitive wholeness of the object that is never won and the subject whose desire is never lost. Desire withholds in order to give, and gives in order to receive. As in the arts, eroticism makes the familiar unfamiliar, and renders desire itself uncanny—its purest state."

Perry Meisel,
Professor of English, New York University

Resisting Nudities

PETER LANG
New York • Washington, D.C./Baltimore • Bern
Frankfurt am Main • Berlin • Brussels • Vienna • Oxford

Florence Dee Boodakian

Resisting Nudities

A Study in the Aesthetics of Eroticism

PETER LANG
New York • Washington, D.C./Baltimore • Bern
Frankfurt am Main • Berlin • Brussels • Vienna • Oxford

Library of Congress Cataloging-in-Publication Data

Boodakian, Florence Dee.
Resisting nudities: A study in the aesthetics of eroticism /
Florence Dee Boodakian.
p. cm.
Includes bibliographical references.
1. Body, Human—Erotic aspects. 2. Nudity—Social aspects.
3. Eroticism in literature. I. Title.
HQ460.B67 306.77—dc22 2008003456
ISBN 978-0-8204-8614-7 (hardcover)
ISBN 978-1-4331-0415-2 (paperback)

Bibliographic information published by **Die Deutsche Bibliothek**.
Die Deutsche Bibliothek lists this publication in the "Deutsche
Nationalbibliografie"; detailed bibliographic data is available
on the Internet at http://dnb.ddb.de/.

Cover design by Eric Fourmestraux

The paper in this book meets the guidelines for permanence and durability
of the Committee on Production Guidelines for Book Longevity
of the Council of Library Resources.

Printed in the United States of America

This book is dedicated to
Jean Luc Nancy's heart donor whoever s/he is

and my father who refuses to stop dancing

Contents

Acknowledgments

b.m.b., pour une tranche de vie érotique
et l'amour doux et fuyant

guy, for his Goldberg Variations,
J-L Nancy's "*Shattered Love*"
and his singular beauty

ricardo, for his incredible vision

emily, for her friendship, tech assistance
and humor

christine & michel, pour 5 rue Parrot

eric, pour sa touche artistique

SUNY at Nassau, for the sabbatical

Introduction

In *Literature and Evil,* Georges Bataille wrote that the base of the erotic instinct could only form itself in the inhuman conditions of a prison; therefore, Marquis de Sade was able to write *Les Cent Vingt Journées de Sodome (One Hundred and Twenty Days of Sodomy)* locked up in the Bastille. This base of eroticism for Bataille and others who followed in his thinking issued forth from a clear and distinct consciousness. In this work, I attempt to reconceive the erotic, its imaginative manifestations and an aesthetic which is ultimately brought about through desire's disruption. Essentially, I take Bataille's notion that "consciousness of desire is hardly accessible: desire alone alters the clarity of consciousness, but it is above all the possibility of satisfaction that suppresses it" and extend it to include the idea that since (using Bataille's construct) the possibility of satisfaction only happens consciously, the erotic (that tenacious piece that is not necessarily equated to satisfaction) may only come into being vis à vis the unconscious. This would follow his idea that Sade, in the Bastille reached a point when the conscious limitations of being were slowly destroyed, so most of Sade's ideas grew out of unconscious mentation. Jean-Luc Nancy's idea of the "Technique of the Present" in *Multiple Arts: The Muses II* as well as Bataille's *Literature and Evil* allow me to establish the poetic nexus found in my own study. Most of the philosophers and theorists, from Bataille, Barthes to Irigaray, Sontag, etc. insist on a clear and distinct consciousness as the base of erotic instinct, but I'm interested here in what develops between "instinct" and "aesthetic" and to what degree unconscious mentation is at the forefront of the erotic. What are the barriers to this unconscious mentation, how do cultural imperatives, especially regarding the nude body play into its development? I use both critical and erotic narratives, some analysis of the nude body/bodies as objects of resistance, applications of Julia Kristeva's *Revolt, She Said* and Jean-Luc Nancy's *Multiple Arts: The Muses II* to construct a parallel grounded in a theory of absence and the psychosocial dynamic of physical and mental surveillance.

At the end of this book, my goal is to move the reader to an understanding of the aesthetics of eroticism as seen through the lens of something that transgresses consciousness, at the least, and moreover create a necessary link between the poetry of *jouissance* and the revolt of body and mind intrinsic to the erotic.

As a disclaimer, I repeat Georges Bataille's words in his Conclusion of *Eroticism* (1957), "Eroticism is the problematic part of ourselves. The specialist can never tackle eroticism." (273) My project here is to revisit the erotic aesthetic in this contemporary moment and contribute to the conversation that has lapsed into a somewhat reductive discourse, for reasons I will try to expose. I use my *resisting nude* and what I establish as the *present Absence of the Other* to offer a new reading of this most intense human moment, the erotic.

I am indebted to all those who have spent years thinking, writing, living this lovely dangerous enigma.

But why does the beautiful never let us go?

—"Paean for Aphrodite"
J.L. Nancy

She pressed her cool mouth against mine. I was in a state of intolerable joy. When her tongue licked mine, it was so wonderful I might have wished my life over.

—*Blue of Noon*
"The Feast of the Dead"
Georges Bataille

He spends a long time kissing me. He's on top of me I feel the weight of his body. The sweetness and warmth of his mouth are intoxicating. . . . .Later, much later, I fall asleep, curled in the hollow of his body. I'm happy and I want to cry.

—*Submission*
Marthe Blau

To touch oneself, to be touched right at oneself, outside oneself, without anything being appropriated. That is writing, love, and sense.

—"Elliptical Sense"
J.L. Nancy

[. . .] no the flesh is never a liar
And the most vicious body remains pure.

Robert Desnos

CHAPTER ONE

The Resisting Nude

The discourse surrounding the body, especially pertaining to power relations has typically taken Foucault's cue straight through to the 21st Century with little distinction between the clothed body and the nude body. Contemporary literary and cultural criticism has certainly raised issues problematicized by resistances to networks of power/knowledge, gender, and the central role of the culturally constructed human body and the power that resides in, around and against it. However, this study moves in a different direction in order to establish a closer inspection of the nude body, its connection to the erotic, revolt and transgression. There is of course an inherent political tension in this analysis, so beginning with some discussion of the nude body in a Foucaultian context is a starting point.

Since Foucault's treatment of the body and sexuality, *bio-power* has generally been understood to denote the "politics of the body." Deleuze, in *Negotiations*[1] identifies Foucault as the thinker who gave birth to *biopolitics,* a post-disciplinary situation where power is thought of in terms of control. The nude body in Western culture is subject to this control, usually involving variations due to gender and or identity parameters. For example, in *Unbearable Weight,* feminist critic, Bordo juggles both Foucault's disciplinary and biopolitical tenets to discuss the objectifying of women's bodies.[2] In other gender studies, such as Judith Butler's *Gender Trouble,* Butler argues, "There is no gender identity behind the expressions of gender; . . . identity is performatively constituted by the 'expressions' that are said to be its results."[3] One's gender is a performance, it is what one does at certain moments, rather than a universal who you are. Stretching Foucault's idea that 'real' identity doesn't exist, that it's just a way of talking about the self, a discourse, Butler confirms the idea that 'identity' is communicated to others during interactions, but this isn't a fixed thing within a person. It is a shifting, evolving temporary construction. Both Foucault's theory and the gender critics reading and extension of Foucault create a basis for some of the inquiries found here. For example, what happens in the case of the nude body when the interaction is erotic, who is reading, interpreting, playing with, controlling, surveilling and/or exerting power over it, how does this body perform differently in relation to the erotic impulse? Is there a possibility for transgression? On the way to understanding the erotic aesthetic, the nude body, its *identity,* coupled with the power relations that surround it and the inevitable revolt it must go through are first steps.

It has been accurately observed that Foucault's *body- as-*

inscribed site trope allows him to consider the discursive conflicts that are part of his analysis, "but the body itself never provides an actual text that is interpreted . . . at no point does [he] actually observe physical bodies."[4] So the physical body is never *seen* by the spectator, Foucault; he makes no interpretation of the body. Using Foucault's framework described by Punday above, we may consider the physical body he does not directly observe, specifically, a kind of "pre-cultural nude", implying the possibility of a nude body existing prior to its interpretation and commodification by the culture to which it belongs, not inscribed with meaning, existing and reacting as its own entity. If a person's identity is simply a way of talking about the self, a discourse, as he suggests and considering his *body-as- inscribed site* trope, how does the pre-cultural nude body fit into this design? It doesn't. The pre-cultural nude body is the physically nude human body, male and female that exists outside a cultural context. It is not seen, it is not the subject of any gaze. The presence of multiple cultures and various power structures in most cultural groupings, along with the consistent presence of a gaze, even if it's only one's own makes the existence of a pre- cultural nude body impossible. If it did exist, how could we explain the disparity between the reactions to Janet Jackson's partial breast exposure during the half time show of the 2004 Super Bowl which caused a huge scandal in the United States, and when televised in the Netherlands and other EU countries for example, it barely elicited any commentary? The cultural component must exist for the nude body to exist, meaning in effect that the *culturally constituted gaze* defines the nude body it sees. So the nude body that could manage to construct or conceive itself outside this cultural backdrop would be an aberration, a resisting mechanism, and as such, this is the only way for it to achieve the status of erotic object (see Chapter 2). We must obliterate culture, and culture here should be read in a broad sense; it refers to that large pool of overseers, norm setters, social and political mandates, that act as 'audience' and in some cases, voyeurs of the nude body. Some might argue that the cultural component described above is always present and that's also true or the resistance would never come about. So, the question we are left with is, without the cultural lens, in whatever form it may take, observing the nude body, interpreting, and creating a frame for it to be critiqued does this nude body then become a subjectless object? Furthermore, does it matter?

Although the discussion of resistance as part of the erotic aesthetic will be picked up in Chapter 6, there is a two-tier resistance at play in this study which can be identified here. One part of the resistance comes from this sociopolitical dimension involving the pre-cultural nude body that is the aberration described above. The second part of this resistance comes from a more profound, internal tension that is key to eroticism and is detailed in Chapter 2. In a sense this pre-cultural nude body that is an aberration of sorts is the precursor to the *resisting nude*. However, the former highlights an externalized resistance while the latter points to an internalized one.

Turning back to some of the socio-political components involving the nude body, the obvious distinction between genital display and the display of the rest of the body raises some interesting points. In Armstrong's "Dispensary"[5] utilized in medical institutions during the late eighteenth century, we find an interesting framework for understanding this phenomenon. It involves Armstrong's findings concerning the 'anatomical atlas':

> The anatomical atlas directs attention to certain struc-

> tures, certain similarities, certain systems and not to others, in so doing forms a set of rules for reading the body, and for making it intelligible . . . the reality of the body is only established by the observing eye that reads it.[6]

Like Armstrong's 'anatomical atlas', attention is pivoted toward certain structures and not toward others. In selecting certain structures, similarities and systems and not others, an immediate distinction is made. "I will look at this, but not that." Therefore, what gets attention seems important, but what doesn't get read also draws attention to itself, simply because it isn't observed. This is key to holistic medicine which the 'anatomical atlas' seems to move against and it is key to seeing the whole nude body. However, the *culturally constituted gaze,* as I'll refer to the gaze of the culture in the larger sense established earlier, directs the viewer/reader away from the genitals; yet, the very attempt to pivot attention in one direction may lead the viewer/reader in the opposite. For example, across most Western cultures, the majority of restrictions are found for exposure of those parts of the human body that put in evidence sexual arousal or sexual dimorphism between male and female adults. However, the attempt to hide the genitals, to suppress a sexual viewing/reading often has the reverse effect. This fact dates back to the Middle Ages when men wore codpieces,[7] later tights and then, tight pants; all these were intended to cover the male genitals but at the same time display them. In the early twentieth century, exposure of male nipples was also considered indecent at some beaches. Ironically, as in the Middle Ages, certain men's bathing suits, while covering the genitals make them quite obvious. This is also the case with the thong, which covers yet simultaneously exposes. The attempt at hiding certain body parts draws increased attention to those parts, in the same way that Armstrong's 'anatomical atlas' pivoted attention toward certain structures and systems to read the body. Here the hiding can be paralleled to Armstrong's pointing, "here this is important", "here are the set of rules for reading this." Rules follow that basically restrict the exposure of genitals. Since the mid-twentieth century, for example, the exposure of genitals is restricted to nudist areas in European countries while topless sunbathing is acceptable on the beach and at outdoor pools. The one exception to the restriction on genital display was Eastern Germany where nude bathing was one of the generally tolerated liberties people could take in the communist GDR.[8] In most Latin cultures, for the most part, genital nudity is not admitted, but women's breasts are now commonly exposed without scandal. In most of the United States, exposure of female nipples (even on mannequins and in lingerie ads) is still not allowed. Public breast- feeding, seen as functional, may be looked upon more mildly, but it is still problematic in most parts of North America. However, Ontario, Canada and New York have legalized the exposure of women's nipples on Equal Protection grounds.

So the mandate to cover the genitals is in fact a restriction on freedom as evidenced by the example of Eastern Germany. The tolerated liberty of exposing genitals freely in public was seen as a relatively minute gesture in the face of a people whose other liberties had been stripped. This would mean that in countries where people have a great of freedom and enjoy many liberties, there is no need to enjoy the right to expose one's genitals in public. This act would be frivolous and marginalized to nudist colonies or the like. The entire nude body becomes less visible in free countries such as the United States. We could also speculate about the correspondence between democracy and the visibility of the nude based on this notion of the liberties granted in a free society. That is, the free democratic society erects itself as a barrier to the nude body and this will later contribute to the impossibility of the erotic in such a place.

In addition, in some cultures, the repeated attempts to cover up the nude body, especially the genitals turns the *culturally constituted gaze* into something more dangerous, more powerful, that is, the development of a taboo and a symbolic system which equates exposed genitalia to the status of medical disorders (localized to a distinct point within the body) of the nineteenth century, that is "dismembered and separated from the rest."[9] The covering in effect acts as a kind of localizing, drawing attention to the genitals rather than away from them. In fact, this is often evident in camera shots that zoom in on tight jean-covered crotches, often "disappearing" the area above the waist. The act of covering itself creates the breaking down of the nude body into pieces, dismembering it, so that what is seen and what is not seen creates the taboo status of the exposed genitalia. As soon as this status is reached, you have what Bernardo Bertolucci describes, "when you go and cover a naked body, then it becomes titillating, obscene." (on *The Dreamers*)[10] It is the covering/clothing that creates the sexual titillation, rather than the fully nude body itself, something fashion designers have been using to their advantage for decades. And while this works to enhance sexual titillation, it has nothing to do with erotic power. As Bataille puts it,

> "The whole business of eroticism is to destroy the self-contained character of the participators as they are in their normal lives. Stripping naked is the decisive action. Naked- ness offers a contrast to self-possession, to discontinuous existence."[11]

The self-possessed removes only what's 'safe', keeping in tact the continuity that comes with sexual titillation that is always ordered by some formula evident in Bertolucci's observation. Not to mention the fact that this marginalization of the genitals via cover-up actually hinders the erotic power of the whole nude body. A dismembered, fragmented corpus cannot carry out its erotic potential.

Another socialized trend worth mentioning here is the naked body of a child in public spaces. In New Zealand, for example, in previous decades, naked children in news and magazines was acceptable; today, it would evoke horror and revulsion. This is due primarily to a shift in social awareness of pedophilia and child porn. In addition, children themselves often absorb parental attitudes about nudity. For example, in the case of children, uninhibited by such concerns, freely walking around nude in Western society, we now see the development of a *auto-surveillance* (discussed in Chapter 5) where children mimic the "covering up" they see in the adult world around them. In cultures where adults freely expose themselves in public places, like on beaches and public pools in most of Europe, Sweden and Norway, children run around nude without the reflex to cover up any part of their bodies. This observation is interesting since it emphasizes how the *culturally constituted gaze* controls the behavior of a child who is still quite young into conceptualizing his/her own nude body in terms of a viewer/reader. However, this study is primarily concerned with the adult nude body.

Sexual difference and the politics of gender of course figure into this discussion of the nude body. The *culturally constituted gaze* takes on a sexualized dimension that is different for men and women. Susan Bordo, in *The Male Body: A New Look at Men in Public and Private,* writes about Sartre and Beauvoir's attitudes about the "Look of the Other."

> Men and women are socially sanctioned to deal with the gaze of the Other in different ways . . . Women learn to anticipate, even play to the sexualizing gaze . . . It's feminine to be on display. Men are taught to be a moving target. Get out of range of those eyes, don't let them catch you-even as the object of their fantasies (or, as Sartre would put it, don't let them "possess", "steal" your freedom).[12]

The power relationship Sartre fears here is obvious. The one who is the subject of the sexualizing gaze has lost his freedom. This phenomenon is certainly not new to women. Women are subject to the sexualizing gaze continuously and the difference with the sexualizing gaze in reference to the nude male body is also obvious.

> The 'full Monty'- the naked penis-is not merely a body part in the movie (hence it doesn't really matter that the film doesn't show it). It's a symbol for male exposure, vulnerability to an evaluation and judgment that women-clothed or naked- experience all the time."[13]

The nude male does suffer from the threat of actual punishment. Display is the punishment. "It seems that it has been intolerable, unthinkable for male evolutionary theorists to imagine the bodies of their male ancestors being on display, sized up, dependent on selection (or rejection) by female hominids."[14] The power relation shifts here, once the nude male is on display, he is totally seen and the bearer of the sexualizing gaze is in control. There is a sexist tenet which underlies Sartre's concern. For example, John Ashbury in New York magazine said of the entire genre of male nude photography, "'Nude women seem to be in their natural state; men, for some reason, merely look undressed . . . When is a nude not a nude? When it is a male.'(Substitute "blacks" and "whites" for "women" and "men" and you'll see how offensive this statement is)."[15] There is an overt desire by heterosexual men to contain the nude male body, to keep it from view and this is directly connected to the power relations in contemporary Western culture. In Foucault's words, keep him out of the "peripheric ring"[16] which can be read as a form of protection for the nude male body, protection against his own surveillance (Is my penis too small? Am I muscular enough? Too fat? . . .) and protection of the power he possesses as the one who is typically the "gazer." Of course, to say that a nude woman is in her "natural" state, implies also that the nude man is in an unnatural state. He is not only "merely undressed" but unnervingly sexual and this is culturally inscribed as unnatural unless referring to a naked homosexual male.

The sexualized male nude body has its cultural roots in gay male aesthetics apparent through the Gucci and Calvin Klein ads Bordo and other gender critics have analyzed. In most cases, homosexual men do not share the heterosexual man's desire to keep the nude male body from view particularly because homosexual men do not fear the loss of power/control by being gazed at, since in most cultures, they do not hold much power in comparison to heterosexual men; they do not bear the traditional role of the male "gazer." This distinction may be directly linked to the homoerotic, that while holding a marginalized site, has a much wider playing field and a freer one.

The propensity to keep the heterosexual nude male body out of the "peripheric ring" might explain the disparity between male and female full frontal nudity in art and media as well. A few general cultural notes are worth mentioning first: the Roman Catholic Church held a "fig- leaf campaign" (the genitals of the nude figures in Michelangelo's paintings in the Vatican's Sistine Chapel were covered with over-painted cloth until the twentieth century) to cover nudity in art, Islam prohibits any illustrations of human beings so the question does not arise there. Scenes of nudity were forbidden in mainstream American films by the Hays Code, until the late 60s-70s when the Motion Picture Association of America film rating system was instituted. Full frontal nudity of both sexes has gained much wider acceptance in European cinema, where the audience perceives non-pornographic nudity as less objectionable than the depiction of excessive violence. Digital imagery may now be used to clothe nude actors avoiding full frontal nudity. "Most movie stars and models are fully dressed even when naked" since they are "purified of 'flaws', all loose skin tightened, armored with implants [and] digitally enhanced."[17] However, the disparity between male and female full frontal nudity is best seen in film. In Hollywood, male frontal nudity is still lingering in its taboo status while the same is not true for women on screen. The full frontal nude shot of Liam Neeson as Alfred Kinsey, in "Kinsey" was cut due to budgetary reasons, a full frontal nudity scene in Colin Farrell's "A Home at the End of the World" was cut after producers thought it was too distracting, in "Bad Education," Gael Garcia Bernal comes close to a full frontal nude shot while preparing for a swim, but he doesn't bare all. In "Sideways", actor M.C. Gainey runs out of the house naked; however, as Jim McBride, who runs mr.skin.com, (that chronicles nude scenes in films), observes, "full frontal male nude scenes appear every few years, and most are limited to art-house releases that are rated R or NC-17."[18] Occasionally, a film such as Larry Clark's "Bully" (2001) is released as unrated or non-mainstream films like the New Zealand film, "Twilight of the Gods" include full frontal male nudity. Certain full-frontal male nudity scenes, like the Robert Deniro, Gerard Depardieu, epileptic hooker one from Bertolucci's European Art House (4+hour film), "Novecento" (Nineteen Hundred) (1977) come to mind, but again these are footnotes in cinema history and as such support the very marginalization described here. Even in countries like France where full frontal nudity enjoys a more relaxed ambience, on French prime time television, full frontal male nudity is not common.

Some film studies critics have argued that a "growing eroticism about the male body that hasn't quite existed before" (Linda Williams, UC-Berkeley) is slowly emerging and the presence of full frontal male nudity "may be a form of rebellion in this post-Janet Jackson environment" (G.A. Foster, University of Nebraska.[19] While Williams and Foster offer possible interpretations of the current status of full frontal male nudity, we cannot ignore the obvious systems of power at play. As Foucault suggests, "power is not possessed; it is exercised. And where there is power, there is always resistance."[20] Consistent with the cultural imperative to keep the heterosexual nude male out of the "peripheric ring" in order to maintain his power as primary "gazer" and keeping his body from being viewed/read, the male-dominated film industry has predominantly followed suit, putting female frontal nudity center stage while the overseer in the tower sits in his protective darkness. The resistance to the power, in the form of more full frontal nudity scenes for men in mainstream film may be slowly emerging, but it is no surprise that Hollywood's old taboo: full frontal male nudity still lingers on considering Foucault's far-sighted biopolitical tenet and the politics of gender which follows from it.

This Foucaultian analysis, in a more contemporary Hollywood frame along with the socio-political aspects considered here are starting points for this study. In order to understand how the nude specifically addressed comes about, and its relation to the erotic aesthetic, the general status of the nude in a cultural context becomes a factor. Culture must be recognized and obliterated which is the impossible tension that gives rise to an aberration and subsequently to the resistance and revolt that evidently follow. So the pre-cultural nude can't exist, but must exist and therefore results in a *resisting nude.*

Culture has not been obliterated from the previous discussion, if anything, it has been invoked, so the structure of the erotic aesthetic proposed must have as one of its possible features, an aberration, a *resisting nude* that takes on the status of erotic object. Since so much research, especially in gender studies, has been done on the "gaze" itself which comes from the subject, we must draw attention away from it and examine the object in its authentic bareness. Essentially the *resisting nude* itself will throughout the course of this study become more and more necessary to the erotic aesthetic. The erotic object will transform into a subjectless object completely bared/d*énudé* while simultaneously being denuded/*l'être-à - nu.*

The next step is to move into the nudity, from the inside as a pathway to understanding what enables the erotic to live. A direct dependence is not at issue, but rather a thread of necessity that must be continually present, always bared while simultaneously being in the nude.

I betrayed my astonishment at a particular caress using the hand as well, whereas I thought the result was supposed to be obtained only by using the mouth [. . .]

Paul Léautaud

CHAPTER TWO

The Bare Necessity of the Kiss

If one could 'enter' nudity as such, this entrance would be marked by a baring, and in the case of the *resisting nude* suggested here that means completely bared/ dénudé while simultaneously being denuded/l'être-à-nu. This level or degree of bareness is the key marker in ascertaining the connection between nudity and the erotic object, as well as how the resulting erotic tension may be understood in terms of poesis or poetry-making picked up in Chapter Three.

So what constitutes the 'bareness' of the erotic object that leads it to its subjectless status postulated earlier? What is the texture of this nudity? How can we view/read it in reference to/or as the erotic object? Keeping in mind that the *resisting nude* is an aberration from the common nude that is always subject to the *culturally constituted gaze,* it has the capacity to become the erotic object or not, unlike the common nude which never reaches the level of 'bareness' constituted here and subsequently is not fully eroticized. This enables the *resisting nude* to take on a variety of forms as the erotic object, including a subjectless object, thereby perpetually transgressing itself. As for example, the egg becomes Simone's eroticized object in Bataille's, *Story of an Eye*:

> Simone settled on the toilet, and we each ate one of the hot eggs with salt. with the three that were left, I softly caressed her body, gliding them between her buttocks and thighs, then I slowly dropped them into the water one by one. Finally, after viewing them for a while, immersed, white, and still hot (this was the first time she was seeing them peeled, that is naked, drowned under her beautiful cunt) . . . [1]

Of course, the eggs later transform into eyes; of the corpse, of the mannequin with a blonde wig and later the bulls 'raw-balls', Granero's dangling right eye and finally the dead priest's eye which Simone calls an egg.[2] One could argue that Simone is the subject gazing in this instance, but the object is never pinned down, or definitely defined by her, even at the end, the eye is not an egg, merely a transgression of itself.

Bareness is the essential component and whether this bareness is borne by a nude body or an egg or an eye, what is important and threatening is that, as Jean-Luc Nancy notes of speech, in particular, poetic utterance, "grappling-with-what-lies-beyond-+[l'être-aux-prises-avec-ce-qui-n'est-pas-soi], . . . there is being that lags behind itself-being in deficit and excess of its own identity and singleness as being."[3] This is how we can begin to understand the bareness found in the erotic aesthetic. That is the bareness of the *resisting nude* implies a resistance to its own indefiniteness, as does the eye in Bataille's *Story of an Eye*. It is the indefiniteness that comes from both deficit and excess of being. The texture of bareness in the first instance extends in both directions simultaneously. Bareness is essentially characterized by both what is missing, absent and the "too much" of being. The simultaneity of these features creates an erotic tension much like the imagistic superimposition of poetry that is further discussed in Chapter 3. However, the parallel I make here finds support in Nancy's contemplation of the following in Michel Deguy's *Gisants*,

> Palms rolling out the pastry of buttocks
> Or the left hand supporting the right breast
> And the thumb softly excising you . . .
> The horizon of thighs displays purple nymphs
> Without an image the sex appears
> And then like a face it is . . . [4]

Nancy suggests that "coincidence as coitus superimposed upon its own image like an ocular-not oracular-bloom"[5] is in fact a proposition that allows for everything and nothing to be revealed simultaneously through

> "the as such of being . . . it is this "as such" itself emerging from its concept and discourse, the "as" as ["en tant que" en tant que] gesture, showing, *deixis,* the presentation of being. And this presentation itself as desire. The showing of being is desire for being, desire for the [as such]of being, for what is desirable is never (naked) being on its own, but the showing in which it comes to be offered up as such"[6]

The gesture, showing, *deixis* is the movement in both poetry and eroticism which keeps the deficit and the excess, if not comparable, then concurrent. As the discussion connecting the poetic and erotic develops later in Chapter 3, I would like to now consider texture, as in the texture of nudity, the texture of bareness, not in stasis, but in movement, as in touching and more specifically, in the gesture of the caress.

The only way to get to the texture of nudity, of bareness is through touch, allegedly the most "superficial of senses."[7] This texture, on the one hand, appears tangible, but it is simultaneously intangible as a result of the absent and excess battling it out in the act of showing. There is both something 'too little' and 'too much' intrinsic to bareness. It feels like something is missing in the midst of sensory overload. This is the tension of resistance embodied in the nude discussed herein which is later transformed into the perceived erotic tension. However, the texture of nudity, of this level of bareness must be considered first, in movement, that is through touch. It is impossible to consider the magnitude of this sense, even for its specific application in this study, without relying on Jean-Luc Nancy's *tactile corpus*[8] and Jacques Derrida's reading and rethinking of Nancy's contemplation of touch. The subtlety of touch for the purposes of this analysis utilizes and redefines their thinking in light of the *resisting nude* and the level of bareness that lead to the erotic. Touch becomes a link between the nude body and erotic potential.

The importance of touch is that it does not necessarily have a limit; it does not begin or end in a fixed sense, but it is "only touch (contact, caress, kiss) [that can] interrupt the mirror reflection in its visual- ocular, optical, or haptical dimension."[9] Since it is a movement, it remains fluid. It communicates something, if we read this in conjunction with Bataille's definition of "stripping naked" . . ."a quest for a possible continuance of being beyond the confines of the self."[10]; there is the invitation to exceed the self through the body, to what, remains unknown.

An early characterization of the caress, by Emmanuel Levinas, "that where I touch without touching, in caressing,"[11] and his observation that

> the caress is a mode of the subject's being, where the subject who is in contact with another goes beyond this contact . . . But what is caressed is not touched, properly speaking. It is not the softness of warmth of the hand given in contact that the caress seeks. This 'not knowing', this fundamental disorder, is the essential[12]

brings us back to Nancy's [l'être-aux-prises-avec-ce-qui-n'est-pas-soi] grappling with what lies beyond which helped define the bareness of the nude earlier and explain its resistance to its own indefiniteness. It is the fundamental disorder of the caress, the not knowing, "an expectation [of something], with no content"[13] that renders the intangibility of the sense itself and this echoes the intangibility of the texture of nudity, of bareness.

This echo, this reverberation of unknowns creates a charged interplay, a place where what lies beyond the indefiniteness meets the contact beyond contact of the caress. And what is this exactly? Using Bataille's terminology, it is where "the rupture of the discontinuous individualities"[14] begins. It is here, that we enter the realm of eroticism. The *resisting nude* is fighting its own indefiniteness, but also desperately needing it to break past itself, its own self-possession. Bataille would call this movement of the *resisting nude* obscene:

> It is a state of communication revealing a quest for possible continuance of being beyond the confines of the self. Bodies open out to a state of continuity through secret

> channels that give us a feeling of obscenity. Obscenity is our name for the uneasiness which upsets the physical state associated with self- possession, with the possession of a recognized and stable individuality.[15]

The recognized stable individuality is upset and this helps us understand the dilemma of touch, especially of the pleasure of the caress: "Where does it come from? From the other or from me? Am I taking it? Am I giving it? Is it the other who gives it to me? Or takes it from me? The time of this pleasure-is it that I am giving it to myself."[16] In the midst of the touch, stable individuality crumbles. The fundamental disorder Levinas describes is vital, but the gender specific assignment he gives as to who does the stroking and who remains 'untouchable' undercuts the inherent disorder of the caress. Levinas's ideas have been reconstituted in post-feminist critiques, such as Luce Irigaray's, particularly his line of thinking that leads to this idea that the one stroking is always masculine and the one stroked (the untouchable) feminine, it is important to consider this earlier use of caress minus Levinas's gender stereotyping or as Irigaray puts it, "his ethical interests [seeing] woman as something to be used by man in his relation with other men."[17]

Derrida observes in Levinas's description, "one has the feeling that she never caresses [and invoking] Jean Luc Nancy's *tactile corpus,* where he speaks of stroking, the latter does not seem to grant a privilege to any one side of sexual difference-and I should rather say sexual difference(s)."[18] Dismissing the idea that Nancy might be neglecting or neutralizing sexual difference, and supporting his analysis, Derrida aptly questions whether we should "presume the sexual identity of the signatory." Assuming the sexual identity of the signatory in the midst of a rupture of discontinuous individualities would dissolve the participation in the erotic, defaulting into a sex game where identities are assigned, completely eradicating the instability of touch and the dangerous ambiguity of the caress.

The texture of nudity, of bareness, in movement through touch is a complex nuanced element of the erotic. Another kind of touch, showing, movement, that is *deixis* of great magnitude is the kiss.

In bareness, as I used to characterize the *resisting nude*, as in Bataille's rupture of discontinuous individualities and Derrida's observation that only "touch (contact, caress, kiss) can interrupt the mirror reflection"[19], we find something that exceeds its own identity, a spilling *over* and maybe *into* another *possibility*. This is a difficult point to pin down because here I am speaking about an intangible space. I will appropriate Derrida's language to call this the place where Touch and Psyche meet[20] or as he reads Nancy's *Corpus*, a "kiss on the eyes . . . a kiss of the eyes on the eyes of the other . . ."[21], but I want to step back here to the more literal kiss on the lips to start with in order to raise a basic philosophical question to further define the *possibility* alluded to above. Whose lips are touching, and whose are being touched? Which lip is touching, and which lip is being touched? Where is the location of the sensation?[22] Perhaps it is more likely that the movement is from the kiss of the eyes on the eyes of the other to the kiss on the lips; however, to speak about the locus of sensation, this movement is examined in reverse, that is, from the more concrete to the abstract, what Nancy reminds us of, "the untouchable of touch"[23] so that we don't over-simplify this sense by giving it some presupposed limit. In this analysis, I do want to maintain Nancy's generous interpretation because the untouchable of touch allows for the *possibility* of moving beyond the edge of the skin and ultimately later, beyond the edge of the self.

Starting where touch makes a concrete contact as in the case of the kiss on the lips, there is still ambiguity. When lips touch, there is no exactitude . . . *where* is the kiss? The sensation creates itself at the moment of contact, and before *and* after. There is a continuum of sensation coming from a point moving toward another point. Is the kiss the sensation? If so, which sensation along the continuum, or all of them simultaneously. The dilemma of the touch is paramount in the kiss where the locus of sensation and ultimately of the pleasure remains unidentifiable. It is in this unidentifiable space that thought, mentation slithers in bouncing off of everything and nothing. I will turn to Nancy's description of *self-touch you* to situate the unique positioning of the kiss.

> *To self-touch you (and not "oneself") [Se toucher toi (et non "soi"]*-or again, identically, *to self-touch skin (and not "oneself")*: such is the thinking that the body always forces to go further, always too far. In truth, it is thought itself which forces itself in this way and dislocates itself: for all the weight, all the gravity of thought-itself a *weighing*-in the end goes toward nothing except consenting to the body and bodies (Exasperated consent).[24]

Nancy's *self-touch you* can be read as "the being of every sense in general, the being-sense of sense, the condition of possibility of sensibility in general, the very form of space and time, and so forth."[25] The condition of possibility of sensibility in general, the very form of space and time rather than a definitive locus of sensation is exactly what is present in the kiss. In fact, if we replace 'touch' with 'kiss' in Nancy's description, we have to self-kiss you (and not "oneself"), to self-kiss skin (and not "oneself"), this is even a more exaggerated form of thought succumbing to the body. Thought ultimately consents to body in the kiss; it is impossible to *think* the kiss. "It is already very difficult to think what happens- and just to think, no doubt, but this may be where "thinking" begins- when a mouth comes in contact with another mouth and when lips, and sometimes tongue and teeth, get mixed up in it."[26] I would extend this to say that the kiss consumes thought and as soon as this happens, "oneself" becomes negligible in the exchange. One cannot kiss (in the mouth-lips-tongue-teeth sense) oneself, and it is this incorporation of the Other in a supremely silent moment when consciousness fails us and we are left with an unconsciousness of continuity between Oneself and the Other, that the erotic is palpable. It is this experience that places the kiss in the most eroticized condition. The mouth-lips-tongue-teeth kiss is the supreme form of touch which cannot be duplicated autonomously as in the case of a masturbatory act or self-arousal.

This extreme condition of the possibility of sensibility shared by the bareness of the nude and the kiss pushes the limits of possibility. The being of every sense in general, the "being-sense of sense" as Derrida reads Nancy's *self-touch you*, opens a pathway to the erotic. The possibility of sensibility is fluid and that could lead the limits of possibility to impossibility, that is, the possibility present in the impossible. We are then left with conditions that permit exceeding identities, breaking down discontinuity, spilling *over* and into *other* possibilities. The invitation to exceed the edge of self through the body is played out in the gesture, the *deixis* of the kiss.

This movement outside the edge of the skin can only be recognized and identified, at least as present, by the Other. Even in the case where the Other is absent, and perhaps more so in this instance, when we actually witness Nancy's "grappling with what lies beyond" [l'être-aux- prises-avec-ce-qui-n'est-pas-soi], the movement beyond the self resonates primarily because, as the *resisting nude* demonstrates, a resistance to its own indefiniteness creates a tension propelling it beyond itself. Toward what? A presence of something else, another possibility that is there even if it's not there. This leads the kisser outward and inward simultaneously. Derrida's use of the sexualized metaphor of day and night to describe the precarious point of possibility ignited by the kiss of the eyes will further clarify this point.

Derrida's reading of the more abstract kiss mentioned earlier, the "kiss on the eyes . . . a kiss of the eyes on the eyes of the other"[27] begins and ends essentially in its relation to the Other, "I love it only inasmuch as it comes to me from the other."[28] A presence of the Other is necessarily there, even when it's not there, when the Other is merely implied as in this case.

The possibility present in the impossibility of this kiss is read by Derrida as, "In the kiss of the eyes it isn't day yet, it isn't night yet. A nightless, dayless point, still. But one day and night themselves are promising each other. One says to the other point-blank: I'm going to give you some. To the point, the break of dawn."[29] In this metaphor, day is present in night and night is present in day even when one is absent from the other. The break of dawn is not dawn, but rather a coming, brought about by the simultaneity of day and night. This point necessitates a presence of the Other in the same way the *resisting nude* battles against it own indefiniteness moving out toward something not itself. In the latter case, the 'something not itself' may or may not be a tangible, present Other. It is simply the result of being completely bared/dénudé while simultaneously being denuded/l'être-à-nu, a bareness in motion that dissolves any definitiveness.

Even in the kiss of the eyes on the eyes of the Other,

as Derrida's metaphor demonstrates, day is only promised and neither day or night can declare themselves fully present. His description invoking a timeless, indefiniteness creates a space for the vibrant promise of continuity between day and night, between eyes and lips.

As we have seen, this indefiniteness is shared by the *resisting nude* through touch, the caress, and most importantly, the kiss. It is further characterized in what I will henceforth refer to as the *present Absence of the Other*. The *present Absence of the Other* is in fact the presence of the Other through its *Absence*. The very "presentness" of *Absence* characterizes the Other, whatever form it may take. The "I'm going to give you some," described earlier in reference to the kiss of the eyes is a promise that can only be rendered by the *present Absence of the Other*. It's out there in the realm of the possible but it is suspended, absent yet unnervingly present. The *present Absence of the Other* is the proof of exceeding being through layers of possibility/ impossibility/possibility. As such, it is the entranceway into the erotic.

As a tool for understanding the scope of the *present Absence of the Other* and its contribution to the erotic aesthetic, the closest parallel to draw upon is found in the poetic. Notwithstanding the esoteric nature of this discourse, poetry-making offers us a way into an aesthetic that has been undercut over time. The imaginative thread that links poesis-making and the erotic for the purposes of this study is primarily based on an extension of Nancy's *access to sense,*[30] especially *making an access be* and Bataille's *access to inner experience,*[31] along with the more critical literary *doctrine of impossibility* found in Bataille's reading of Baudelaire, and Sartre on Baudlaire in *Literature and Evil*. In bringing these two thinkers together in the following chapter, *the present Absence of the Other,* that is, the proof of exceeding being may be read as an access to the erotic.

I only remember the scent of pine trees [. . .],
the taste of salt under my tongue [. . .]
And the imminence, the unbearable imminence of pleasure
Suspended at a point in time that was vaster than eternity

Maïssa Bey

CHAPTER THREE

The Sense Connection:

Poesis Making and the Erotic

The proximity of the publication of Bataille's *Eroticism* and *Literature and Evil,* in France, in 1957 by Les Editions de Minuit and Editions Gallimard respectively mark a convergence that I would make note of here, in addition to Jean Luc Nancy's later work, "Making Poetry" in *Muses II* first published as "Faire, la poésie", in *Nous avons voué vie a des signes*[1] that come together to demonstrate the erotic as a cerebral dance played out by the senses. The erotic is established through the intersection of sense and imagination, and in both this way and its representative function shares an interesting correspondence to poesis making.

Bataille uses the same line in both *Eroticism* and *Literature and Evil,* "Eroticism, it may be said, is assenting to life up to the point of death."[2] He uses this same description writing about Emily Brontë's life and work in *Literature and Evil*, "I believe eroticism to be the approval of life, up until death."[3] The content of these lines, for the moment, is not necessarily key, but rather the timely train of thought carried by Bataille through his analysis of both literature, especially poetry, (as in the case of Baudelaire) and eroticism. It is no surprise that in the forward to *Eroticism*, Bataille pays homage to Michel Leiris[4] and his work, *Le Miroir de la Tauromachie*, "I should like to mention here that my own endeavors have been preceded by *Le Miroir de la Tauromachie* by Michel Leiris, in which eroticism is envisaged as an experience wedded to life itself; not as an object of scientific study, but more deeply, as an object of passion and poetic contemplation. This book [*Eroticism*] is dedicated to Michel Leiris particularly because of this book of his . . . written just before the war.[5] The analysis Bataille offers on the *doctrine of impossibility* (to unite being and existence objectively) in Baudelaire's poetry is especially relevant to his own work on eroticism written closely together. Interestingly enough, there seems to be a moment of subtle intersection that hung in a philosophical limbo later developed in certain theories of poesis making like we find in Nancy's work.

The exercise of viewing eroticism through poesis making involves, in addition to the doctrine of impossibility, the access to sense/making an access be, access to inner experience and the technique of the present, essentially a merging of Bataille and Nancy's theories in combination with the model of the *resisting nude* and the *present Absence of the Other* to arrive at a new reading of the aesthetics of eroticism.

In constructing the model of the *resisting nude* and in contemplating touch, we saw that what is missing, absent and the "too much" of being characterizing bareness result in the kind of tension Nancy uses to describe speech, especially poetic speech. That is, the bareness (posed) by the nude resisting its own indefiniteness is like "grappling-with-what lies beyond-+ [l'être-aux-prises-avec- ce-qui-n'est-pas-soi] . . . that is being that lags behind itself-being in deficit and in excess of its own identity and singleness of being."[6] So, this *resisting nude,* in resisting its own indefiniteness in fact resists its inherent capacity to exceed itself. This is the same extension we find in touch, especially in the case of the kiss where the extension *must* occur. In a sense, the postulated resisting nude is both object and subject. It may, at times, appear to be a "subjectless object" as referred to in Chapter 1. Here is where the most obvious, yet simultaneously elusive poetic parallel exists.

> But whichever way we look at it the synthesis of the unchangeable and the perishable, of being and existence, of the object and subject, which poetry seeks, is an ultimate definition of poetry. It limits it and transposes it into the realm of the impossible and unsatisfiable.[7]

In the erotic, in the movement of extension, object and subject can appear to come together as noted; the acutely indefinite posture of the *resisting nude* attempts to unite being and existence and in striving toward this impossibility becomes more and more eroticized. The poetic and the erotic share the same desire: the desire for the impossible.

Bataille's treatment of the doctrine of impossibility or as he calls it, Baudelaire's *statue of the impossible* (la statue de l'impossible) basically suggests that the poetic process wants the object to become the subject, the subject the object or that in its quest for the identity of reflected things and the consciousness which reflects them, wants the impossible. Yet the means of avoiding reduction to the reflection of things constitutes a desire for the impossible.[8] Baudelaire kept himself in a permanent state of unsatisfaction, by writing poetry, especially in the case of *Les Fleurs du Mal,* and an agonizing attraction to an unsatisfactory life was an externalization of the poetic process he created. Baudelaire and his poetry are highly eroticized keeping the satisfaction in abeyance while perpetually remaining in a state of desire. The *resisting nude* shares a similar dilemma in its shift between subject and object desperately attempting to unite being and existence, eroticizing itself in the process. As it resists its own indefiniteness, it desires what is out there, that is, the consciousness and the reflected thing itself, a complete impossibility.

In *Eroticism,* Bataille suggests that the breaking down of discontinuous selves and moving toward continuity is a necessary part of the erotic.[9] Interesting though, it is at this moment when he also admitted that poetry could help him explain the breakdown and fusion present in the erotic, but resolved that he would not talk about poetry.[10] This impossible necessity is alive though in the poetic prose of his *Story of the Eye* in the sublimely erotic bicycle ride the narrator and Simone take to flee the château. Since their clothing was lost, they are both nude and exhausted, pedaling frantically with 10 kilometers left to their destination. He has an erection watching his friend pedal in front of him and realizes it's difficult for Simone to see him, "[y]et I felt I could see her eyes, aglow in the darkness, peer back constantly, no matter how fatigued, at this breaking point of my body, and I realized she was jerking off more and more vehemently on the seat, which was pincered between her buttocks. Like myself, she had not yet drained the tempest . . ."[11]

Throughout Bataille's *Story of an Eye,* the reiteration of his analysis of Baudelaire is clearly apparent. "The essence of Baudelaire's poetry is to affect, at the cost of an agonizing tension, the fusion between the subject (immanence) and those objects which lose themselves both in order to cause anguish and to reflect it."[12] In the erotic scene above, Simone's nude body is hurled upon an embankment as "she was literally torn away by joy" and the narrator himself is "overcome with bloody spasms, my lower lip drooling and my teeth bared like a leering moron"[13] as he attempts to bring her back to consciousness. The narrator clearly exhibits a painful, yet pleasurable tension in his own body in correspondence to Simone's *jouissance*. What Bataille conveys here that goes beyond simple sexual titillation is a resistance stemming from the narrator's attempt to counteract his own indefiniteness, to carry out the act of bringing Simone back to consciousness while his own senses are reeling; he wants to fall asleep beside her and eventually does, losing consciousness himself, diluting the line between subject and object, temporarily. The moment when the impossibilities shared by poesis making and the erotic become possible is found in a state of unconsciousness or death. That explains the narrator's thought, "And it struck me that death was the sole outcome of my erection, and if Simone and I were killed, then the universe of our unbearable personal vision was certain to be replaced by the pure stars. . ."[14] The synthesis of being and existence occurs when they cancel each other out.

Neither the erotic nor poetry is a definitive act, but rather propositions. Something is being posited in each case, but also not, and in the case of poetry, Nancy identifies this as the access to sense or making an access be. As such, "the selfsame thing that is both abolished and posited is the access to sense. Access is unmade as passage, process, aim, and path, as approach and approximation. It is posited as exactitude and as disposition, as presentation."[15] Essentially access to sense keeps the continuum of possibility open. Nancy points out that when access appears, it is as if it was always there, precisely because the posing and deposing both come into play. "When saying is making, and making saying, in the same way that one says "making love" which is making nothing, but making an access be. Making or letting: simply posing, deposing exactly."[16] Definitive lines of beginning and end are not evident; in the gesture of posing, putting forth something is also the movement of deposing, necessary for art, and especially poetry, and clearly key to eroticism. Deposing, *déposer,* to lay down or aside; to divert oneself of; to depose, to deposit, to lodge is the other end of posing. The making of poetry is always in motion, in act, in Nancy's words, "in the immemorial pluperfect."[17] Therefore speaking about the aesthetics of poetry is tricky and often can only be read as an afterthought to making that is in action long before it can be identified.

The making of the erotic follows a similar path as the *present Absence of the Other* identified in Chapter 2 points to, that is, the "presentness" of *Absence* characterizes the Other, whatever form it may take. It is the absent, yet unnervingly present proof of exceeding being through layers of possibility/ impossibility/ possibility. I am not suggesting that Nancy's description of deposing can be aligned to the *present Absence of the Other,* but rather that the outcomes of both create a space that allows the making to occur, that permits an access. In addition, they share *presence;*

> Presence is not a quality or a property of the thing. Presence is the act by which the thing is put forward: praeest. From 'already no longer' to 'not yet' is a passage without pause, a step not taken, neither disposed or exposed, inexposable, only and ceaselessly deposing all things.[18]

This brings us back to the same precarious point of possibility in the breaking of dawn metaphor used in relation to the kiss. The 'already no longer' and 'not yet' like the presence of day in night and night in day and like the *resisting nude* battling against its own indefiniteness moving out toward something not itself ultimately defined by the *present Absence of the Other* all point to the simultaneity of possibility and impossibility.

The impulse to steer away from a discussion of poetry that Bataille expressed in *Eroticism* is of course, ironic, both from aesthetic and philosophical vantage points. The bridge composed of the *resisting nude* and the *present Absence of the Other* that I have attempted to create here between Bataille and Nancy brings to surface the depth of this Bataillian irony.

Bataille's discussion of the access to inner experience contributes a further dimension. Much of Bataille's emphasis in *Inner Experience* revolves around states of ecstasy and mysticism that often lead him to parallel critiques of religion and eroticism. The connection to realms of impossibility and the unknowable create an interesting point of contemplation for this study. He writes, "the inner experience is throughout an experience of the *impossible* (the impossible being both that which we experience and that which constitutes the experience)"[19] Bataille can only examine eroticism from this vantage point as well; without entrance into the realm of inner experience, one cannot begin to come to terms with the erotic. "Human eroticism differs from animal sexuality precisely in this, that it calls inner life into play."[20] The inner experience that has virtually no outside manifestation other than how it may be objectified is problematic; therefore, access to inner experience becomes as necessary as access to sense, clearly making an access be, gaining entrance into the experience of the impossible. This is what poetry and the erotic attempt to do. As Nancy argues, without this access, poetry doesn't occur. Perhaps something resembling poetry is elicited, but it isn't poetry. In the same way, without access into the realm of inner experience, into this experience of the impossible, the result is basic sexuality, not eroticism.

This obvious parallel between poesis making and the erotic leads to a discussion of the distinctions between the sexual, the pornographic and the erotic. Laurence R. Scheler and Jean Francois Fourny, in "Writing After the Erotic" raise the question of whether distinctions such as those made between the sexual, the erotic and the pornographic in literature are passé symptoms of an advanced state of sybaritism? Or are they a new way to rethink the social, the personal and even the political in a secular, post-modern world that has, at least in much of Western Europe, abandoned its reliance on grand narratives.[21] I would argue that the kind of difference that becomes self-evident in the poesis making parallel above precludes tossing all distinctions aside; however, they must be reconsidered in light of the contemporary moment.

Hours pass, one or two minutes pass, time is a beating full of froth and whip lashes.

Julio Cortázar

CHAPTER FOUR

Porn and the Erotic: A Border of Impossibility

While poesis making continues to thrive even in industrialized cultures where all kinds of 21st century digital distractions threaten the access to sense, the erotic is riding on a much more problematic tightrope. Some of the distinctions that can't be readily dismissed may explain why. However, I would further suggest that eroticism is close to and almost extinct in certain Western cultures, and especially in the United States of America. The current political and social climate can't sustain it.

The vanishing distinctions that Scheler and Fourny allude to are a reflection of an attempt to recast the sexual in a new kind of discourse, for example, in France there is a current movement to rethink and rewrite some of the previous borders such as those between porn and the erotic, between hetero and homosexuality, and between the literary and non-literary, as evidenced by a recent copy of *le magazine littéraire "les enfers du sexe de Sade à Houellebecq."* There is certainly evidence of a change echoed in the questions raised by Scheler and Fourny in "Writing After the Erotic" but what may look like a vanishing of borders may in fact be a reduction or distillation of the erotic.

Mario Vargas Llosa wrote the following lines in *The Notebooks of Don Rigoberto,* 1997, in "Don Rigoberto's Letter to the Reader of Playboy, or "A Brief Treatise on Aesthetics":

> After traveling a long, long road whose backbone, in a sense, was the progressive distillation of erotic play by an unexpected route-the permissive society, the tolerant culture-we have returned to our ancestral starting point: lovemaking has again become physical, semi-public, thoughtless gymnastics performed to the rhythm of stimuli created not by the unconscious mind and the soul but by market analysts, stimuli as stupid as the false cow's vagina passed under the noses of stabled bulls to make them ejaculate so that their semen can be collected and used for artificial insemination.
>
> Go on, buy and read your latest Playboy, you living suicide, and bring your grain of sand to the creation of that world of ejaculating male and female eunuchs where imagination and secret phantoms will vanish as the pillars of love. For my part, I am going to make love to the Queen of Sheba and Cleopatra, both at the same time, in a play whose script I do not intend to share with anyone, least of all you.[1]

The under-played erotic imagination he describes has progressed down a contemporary path intensified by Reality

TV, the politicized demonization of the emancipation of eroticism from both sexual reproduction and love and an environment of obsessive surveillance coming from both external and internal forces. While Llosa's complaint may already seem outdated, his mourning of the deadening of the unconscious mind and imagination are even more fitting now. The access into the realm of inner experience, making an access be, entering into the experience of the impossible are thwarted in contemporary society for several reasons.

As we saw, the erotic requires a movement towards the limits of one's subjectivity and humanity, and as Bataille suggests, a transgression that dissolves the rational world but is always transitory.[2] It is not fixed in place, belonging only to the transient present finding little breathing space in a culture that demands absolutes ("you're either with us or against us"), holds *reality* without transformation or transgression as an art form, promotes a fixed future through the socio-political support of the institution of marriage and sexual reproduction within those parameters, and has created an atmosphere of consistent surveillance to pin down anything before it happens. This absolute certainty is in direct opposition to the erotic that remains a site of conflict and contradiction while actually functioning as a liberating force. Furthermore, absolute certainty cancels out the experience of the impossible that we have seen is necessary to the erotic. Contradiction is the evidence of vacillation between possibility and impossibility. In Audre Lorde's *Zami*, her erotic fictionalized autobiography, and in "Uses of the Erotic as Power," Lorde deconstructs the objectified image of the exotically garbed Caribbean woman away from its conscription of human erotic potential servicing the phallocentric/colonialist/imperialist gaze.[3] She has thereby expanded the range of erotic vision while adding contradictory implications for the discourse of the erotic.

> I took a ripe avocado and rolled it between my hands until the skin became a green case for the soft mashed fruit inside, hard pit at the core. I rose from kissing your mouth to nibble a hole in the fruit near the navel stalk, squeezed the pale yellow-green fruit juice in thin ritual lines back and forth over and around your coconut brown belly. The oil and sweat from our bodies kept the fruit liquid, and I massaged it over your breasts until your brownness shone like a light through a veil of the palest green avocado, a mantle of goddess pear that I slowly licked from your skin.[4]

The trait exposed in Lorde's erotic work points to what Barbara Christian characterized as typically Caribbean, that is, "that people create their own forms."[5] Therefore, Lorde's Caribbean sensibility is one demonstration of the erotic as a transformative source which negates the absolutism described above, especially that of surveillance that inherently requires a fixed, or at best, non-contradictory subject.

The continuum of possibility/impossibility/possibility necessary to poesis making and the erotic eliminates absolutes leaving behind very little certainty. In the *Metaphor of the Eye,* Barthes picks up on Bataille's play with the continuum of possibility in *Story of an Eye* "At one point during the sexual activity something in particular took place . . . and then the activity moved on to something else."[6] This something else cannot be registered because it is outside of the realm of "knowable" possibility, an impossibility. It has transgressed a line that can be named and I would argue that this transgression is a distinction unique to the erotic. Transformation and transgression are likened to that unknown other, to that indefinite space, to a "counter-division of objects, usages, meanings, spaces and properties that is eroticism itself."[7] Of course, pornography is marketed as transgressive because that enables its titillating appeal, but it is nevertheless a controlled product employing the instruments of a staged performance even if the acts themselves unfold in one's own bedroom. Susan Sontag, long before the Web-enhanced porn banquet of the 21st century treated Bataille's *Story of an Eye* as 'meta-pornography' in her essay "The Pornographic Imagination." As early as 1969, she engaged in the blurring of boundaries Scheler and Fourny point to in "Writing After the Erotic." However, she was still compelled to call it something other than pornography. Sontag cleverly confines her discussion of pornographic imagination to the realm of consciousness, albeit the "extreme forms of consciousness that transcend social personality or psychological individuality."[8] Bataille, on the other hand, in *Eroticism*, describes a loss of consciousness which exposes the "inner experience" that necessitates the convergence of the erotic and death.[9] It is this access into the realm of inner experience, into this experience of the impossible that must occur for the erotic to exist. Sontag implies the ambiguity around these distinctions; "the question is not whether consciousness or whether knowledge, but the quality of the consciousness and of the knowledge."[10] She finds the source of pornographic imagination in consciousness, yet she uses *Story of an Eye* which follows Bataille's own prescription for the unconscious imperative proclaimed by the erotic and death described in Eroticism and evident in his narrative, *Story of an Eye*.

> Granero was thrown back by the bull and wedged against the balustrade; the horns struck the balustrade three times at full speed; at the third blow, one horn plunged into the right eye and through the head. A shriek of unmeasured horror coincided with a brief orgasm for Simone, who

> was lifted up from the stone seat only to be flung back with a bleeding nose, under a blinding sun; men rushed over to haul away Granero's body, the right eye dangling from the head.[11]

The conscious depersonalization of sex organs used to arouse in pornography is quite the opposite of Bataille's project; he personalizes the objects including the sex organs. In addition, as Sontag suggests, "pornography becomes a group pathology"[12] whereas eroticism involves individualized acts that converge. In the bull scene above, the simultaneity of Granero's impending death and Simone's orgasm signal a convergence marked by an impossible possibility.

Pornography operates within a sphere that almost necessitates the spectator, the watching; the surveillance is part of the titillation. Of course, this is not entirely absent from the erotic; however, the transgression required of eroticism, as Bataille demonstrates, is dangerous to the extent that it creates a landscape which cannot be surveilled; it is not the result of consciousness, but of a drop out of the "habitually desired balances"[13] (which can be easily monitored and surveilled) into the "equally habitually desired transgression of those habits that makes human embodiment so uncanny."[14] This distinction takes on a new dimension when read against the current cultural imperative issuing from a surveilled global landscape.

The tendency of eroticism is to supercede a coded ready-made format as Sontag suggests of pornographic imagination, that is, a "[preference] for ready-made conventions of character, setting and action. Pornography is a theatre of types, never individuals."[15] On the other hand, the erotic work, *The Story of O*, by Pauline Réage for example, does not contain a "pure, impersonal" sexual act-there are no 'pure' sexual encounters possible. "O" reacts in different ways to

Rene, Sir Stephen, etc. Everything is personalized, one person is not exchangeable with another, as Sontag suggests is common to the pornographic.[16] Exchangeability is not possible in the erotic because of the *present Absence of the Other* and the subsequent doctrine of impossibility described earlier.

The existence of these conventions described by Sontag makes a code possible. However, eroticism necessitates individual ego to transcend individual direction and to create a space for what Bataille called, "self-violation and unthinkable societal transgression."[17] Obviously, the erotic exists beyond a code. It is accessible through the access to inner experience made possible through the experience of the impossible identifiable through a loss of consciousness. The poetic endeavor and the *resisting nude's* dilemma to unite being and existence, to resist its own indefiniteness mentioned earlier in this chapter, all point to a striving toward an impossibility and in the case of the latter, in doing so, becomes more and more eroticized.

In a "Preface to Transgression", Foucault wrote that it is not so much that sex finds a language of its own in pornography but rather that language does violence to sex, limiting its capacity to point beyond itself.[18] Sexual discourse is a form of exposure and in itself simply another code. Overexposing the "displayed" experience through sexual discourse in whatever form it appears in offers a resting place for "reasonable" desire. This desire dwells in the conscious mind, can be coded and read, nourished by the boundaries of the language which communicates it. The erotic involves something else, "an overpowering return to a lost continuity as the eruption of elements opposed to reason, as a breathtaking act of self-de-limiting,"[19] Reliance on a radical loss of consciousness necessary to bring the erotic to surface cannot be aligned to carefully manufactured, coded porn.

It is the act of self-de-limiting or as Nancy explained, the "grappling-with-what lies beyond-+ [l'être-aux-prises- avec-ce-qui-n'est-pas-soi] . . . being that lags behind itself- being in deficit and in excess of its own identity and singleness of being" (see Ch.3 n6) that is key. The question, of course, remains, at what point does this act of self-de-limiting occur? Or can it, here? Now? This is why the *resisting nude* postulated herein remains in a state of resistance. Furthermore, opposing reason, allowing for unspeakable self-violation and unthinkable societal transgression do not lend themselves easily into contemporary culture laden with surveilled landscapes and escalating fear.

One might argue that unthinkable societal transgression is present, here is one brief example, in the nymphet eroticism that is part of the pop culture repertoire; however, as Richard Goldstein points out in his article, "The Double Standard", "straight guys' eroticizing of teenagers is a mainstay of pop culture, so why is it a disease when gay guys do the same thing?" The culture pathologizes gay men's lust, so parameters are set up; nymphet eroticism is "reasonable" within the heterosexual community, step out of that domain and we have *ephebophilia* (the disease characterized by the seduction of teenage boys by gay men).[20] A breath-taking act of self-de-limiting within a culture that imposes standards of acceptable behavior, and especially reasonable desire, while simultaneously posing as the "permissive society", the "tolerant culture" has very little chance of coming into being.

The heightened popularity of Reality TV to a pseudo- art form suggests another block in that it masks as an entrance into the private arena. It is a staged private arena suffering from a lack of transformation or transgression and merely parades its participants around in their sexy lingerie. The interesting component of Reality TV, as well as Web porn,

YouTube, My Space, etc. is that the arena has moved expansively; however, the erotic has slowly faded out. A more restrictive environment has surfaced. These spaces heighten consciousness and bring participators and spectators into a "safe" place. There is no drop out of the "habitually desired balances" Bataille speaks of, no movement into the experience of the impossible, no present loss of consciousness, no grappling beyond the edges of being à la Nancy.

In fact, the societal transgression Bataille makes reference to counteracts the "safe" place. It is challenged by the disruption of the obscene embodying the gratifications of desire and death. Restlessness caused by intense sexual desire and intense pain as one comes toward death rock the foundation of security. This is a scary proposition in a culture which has become accustomed to its staged sexual play and its packaged porn treats that require total consciousness, vigilance, and in some case, a credit card. Nancy observed that poetry declined "in a period when language was strictly controlled and exact."[21] Likewise, the decline of the erotic occurs under these same conditions. The difference in the case of the erotic is that the new spaces that have emerged distract us away from the fact that control and exactness collapse all opportunity for self-de-limiting. What we have before us is the appearance of landscapes of freedom, but the unconscious terrain remains hidden away and Mario Vargas Llosa's characterization of the progressive distillation of the the erotic rings truer now than ever in a global environment where surveillance is heavy. The goal of surveillance is to pin down anything before it happens and eroticism operates as a state of arousal and anticipation, it's transitory, and forever lingers in that frightening space of im/possibility.

I bit his chest all over: electric currents run through my tongue, and gums [. . .] There was something like a burst of flame and my brain reached an orgasm, alone and silent, magnificently alone.

Alina Reyes

CHAPTER FIVE

Surveilled Body/ Surveilled Mind

The philosophical implications found in chapters three and four concerning eroticism are directly linked to the psychosocial dynamic of physical and mental surveillance. Considering our contemporary cultural landscape, we are surrounded by a new phenomenon of surveillance that has crept in from the political to the personal. In the last seven years, since September 11, 2001, everyone is watching everyone. Surveillance has become the code word of a so-called free society. Within a surveilled society, individuals begin to watch themselves too. *Auto-surveillance* is a direct manifestation of the reflex to watch the other being watched.

Sweeping past the Orwellian eye, we now engage in *auto-surveillance* as a by-product of an overly protected, paranoid, socio-political climate. Essentially the Other is irrelevant and the subject/object lines are now fuzzy. *Auto-surveillance* is evidence that being watched over time surpasses its physical bounds and slowly invades mentation. How does this happen and what are its implications for the erotic?

A recasting of Bentham's Panopticon and Armstrong's Dispensary, both found in Foucault's study of the reform of medical institutions in the late eighteenth century in an attempt to discover "how the medical gaze was institutionalized [and] how it was effectively inscribed in social space"[1] offers some interesting aspects for consideration in this new culture of heightened *auto-surveillance*. Through it, we can discover how surveillance may be hindering the erotic aesthetic in contemporary Western culture.

Bentham's 'device', the principle of the Pantopticon worked as follows:

> A perimeter building in the form of a ring. At the center of this, a tower, pierced by large windows opening on to the inner face of the ring. The outer building is divided into cells each of which traverses the whole thickness of the building. These cells have two windows, one opening on the inside, facing the windows of the central tower, the other, outer one allowing daylight to pass through the whole cell. All that is then needed is to put an overseer in the tower and place in each of the cells a lunatic, a patient, a convict, or a schoolboy. The back lighting enables one to pick out from the central tower the little captive silhouettes in the ring of cells. In short, the principle of the dungeon is reversed; daylight and the overseer's gaze capture the inmate more effectively than darkness, which afforded after all a sort of protection.[2]

The Panopticon and later Armstrong's "Dispensary" created spaces for surveillance. "Where Panopticism had in effect offered to the observing eye that which was an enclosed space, the school, the hospital or the prison, the Dispensary through its juxtaposition of observation and community replaced enclosed physical space with an open social domain."[3]

The prisoner in the scenario Foucault picks up from Bentham's Panopticon isn't aware that he may be watched, "he becomes the principle of his own subjection."[4] He surveys himself since he is not certain that he is being watched, but the possibility is there, whether or not an overseer is actually in the tower. Looking back to some of the earlier observations in the first chapter regarding the nude body, *auto-surveillance* plays out in a similar way. For example, on American beaches, even after lifeguards, beach patrol officials and others have left the beach in the heat of midsummer, a lone adult does not undress down to nudity. In this case, the adult man or woman who puts perfectly dry clothes over a soaking wet bathing suit isn't being watched. In the case of the American beach, it would matter that no one is around to watch, since undressing and nudity are prohibited on most public beaches in the country. The *auto-surveillance* here may reside in *gymnophobia,* the fear of being (or seeing others) naked, but that is too narrow an interpretation for what is going on in this example. The stimulus for the fear is the *culturally constituted gaze* that remains omnipresent. The proof of this is that in many European countries where the nude body is not prohibited from public display on most beaches, adult men and women swim and sunbathe nude and undress and redress on the beach whether or not they are alone or in the company of others. The *culturally constituted gaze* is more than a cultural lens because it has this *auto-surveillance* aspect that implies a controlling of the nude body for culturally determined reasons that later become interiorized. The omnipresent power of the *cultur-*

ally constituted gaze converts to *auto-surveillance* making the Other irrelevant. Power is exercised in terms of control à la Foucault since the culturally constituted gaze functions also as, "an inspecting gaze, a gaze which each individual under its weight will end by interiorisation to the point that he is his own overseer, each individual exercizing this surveillance over, and against, himself."[5] So that the only possible resistance to this interiorisation is exteriorisation, the nude body itself in public and private.

The dissociation of the see/being seen dyad characteristic of the Panopticon is revisited in Armstong's "Dispensary". Unlike the Panopticon, which was a device for 'monitoring and constituting bodies', the Dispensary is a mechanism for surveying, and thereby rendering problematic, particular relationships between those same bodies.[6] The metaphor of the Dispensary is even more important to eroticism because it is an open social domain. It operates as a 'new perceptual structure' and is at the core of what Foucault called "surveillance medicine" which concerns how the gaze explores localized pathologies and revolves around the observation of social networks and relationships.[7] The open social domain of global digital media including some of the those mentioned earlier, such as the internet, provides this community aspect of the Dispensary to contemporary culture. The isolationism inherent in the Pantopticon is replaced by an open social domain where surveillance is much more palpable. The *present Absence of the Other* can't exist since the Other is never absent. Since this *present Absence of the Other* has previously been established as a vital component of the making of the erotic, it is clear that the open social domain acts as a further obstacle.

The other Foucaultian observation, in relation to surveillance medicine, that is relative to this discussion is "the recruitment of the 'active patient'-self-observing actors, monitoring their own body and 'lifestyle' for signs of abnormality and deviation from the role of responsible individual citizen and urged into compliance by the medicalisation of everyday life."[8] Of course, the exercise of power here is continuous and cost-effective as Foucault points to, and it also inherently masks itself, so the *auto-surveillance* is never really articulated and therefore never read for what it is, an interiorisation of external control.

If as we have seen, the erotic necessitates the following: an access, primarily to sense and inner experience, the *present Absence of the Other*, that is proof of exceeding being (see Chapter 3) and as Bataille has observed, it "always entails a breaking down of established patterns, the patterns of the regulated social order basic to our discontinuous mode of existence as defined and separate individuals"[9], then how can it come into being in an environment where *auto-surveillance* is active? The aspect of exceeding being is impossible since the Other is always present, watching, controlling, bringing order into place; there is never an absence, a solitude of being since under the grip of *auto-surveillance,* the gaze has turned inward and the Other becomes the Subject. In addition, the access to sense and inner experience requires a disruption of the "habitually desired balances" (see Ch.4 n13) as Bataille illustrates and this means deviation and transgression, not fulfilling the 'active patient'/'responsible citizen' Foucaultian model. In the previous chapter, I argued that the erotic cannot be surveilled and the inverse of that is simply that under the fog of *auto-surveillance* it simply doesn't appear. It cannot be surveilled because it never comes to surface. So it is protected then and can live freely? Not really, not without resistance of the kind we have seen exhibited by the *resisting nude,* Nancy's posing/deposing and Bataille's loss of conscious control.

The 'new perceptual structure' of the Dispensary of the eighteenth century can be paralleled to the open social domain of global digital media including the internet. The unique twenty-first century dimension is that this is where the deviant and uncontrolled irresponsible citizen, once exposed, finds him/herself eternally there. The 'new perceptual structure' operates as a physical domain, but more importantly it is a mental space. A self- monitoring occurs and despite the physical openness of the space, the participator or spectator internalizes the systematic restrictions in the role of responsible citizen. Even the Web porn mentioned in the previous chapter operates through a pornographic imagination that Sontag characterized as patterned, coded and a 'theatre of types' so within that cyber world, parameters are set, rules clear with little transgression.

The same thing that urges the 'good patient/the responsible citizen' into compliance in Armstrong's Dispensary motivates the contemporary citizen, that is the 'medicalisation' of everyday life. The imperative driving both is to *make it normal* and be attentive to signs of abnormality and deviation. We are now aided in this project by the open surveillance emitted from the internet, digital tracking, heightened security, and in the United States, the Patriot Act.

This kind of surveillance has diluted the efficacy of the aesthetics of the gaze. Here, I am referring to the gaze belonging to beauty and the impulse of art (like poetry) and eroticism, not that which belongs to culture per say. In his discussion of the "Technique of the Present," Nancy examines this gaze in a way that is relative to the analysis here.

> Vision sees the present, but only because the present opens the eye, and disposes it to its presentation. The present is this disposition which exposes the thing to its vision, that is, to its desire to be open to itself . . . It is the

> thing cut by the present which distances it from itself-by the plane that colors and exhibits it.
>
> It is the thing as this thing. The thing at an encounter with itself. My gaze is this encounter. It is not the representation of the thing, nor am I the subject of a representation nor of an intention. I am the staring present of the presentation . . . Poetry is the productive technique of this end. For what one wants to feel is nothing but this: this end, each time, this overflowing of the present in any present whatsoever. That's what one strives to fabricate, behind the immense and perceptual flowing of presence. But where, behind where, and when? There where time opens . . . [10]

The desire to be open to itself is a starting point granted by the present in the poetic, and the same is true of the erotic. Earlier, in relation to the *resisting nude,* we saw how in resisting its own indefiniteness, it desires what is out there, that is, the consciousness and the reflected thing itself, a complete impossibility. The 'thing as the thing' that Nancy describes above creates a possibility within this impossibility with the gaze as encounter, but the gaze is not a representation of the thing. 'I am the staring present of the presentation.' It is the encounter with itself at a place where time opens up, at this point of a over-flowing present. The desire to feel this end each time in the present as Nancy describes is a frightening and seductive call. It implies a mental space in direct opposition to that of the 'new perceptual structure' described above. It is the neither disposed or exposed, open space where the simultaneity of possibility and impossibility is possible. This gaze belonging to the beauty of the never-ending present makes this space visible. The internalized gaze in the Foucaultian analysis should be differentiated from this later gaze Nancy is referring to, that is, the one bearing the aesthetic presence. The internalized gaze is an external one turned inward while this one begins and continues, with no perceivable end, internally. If *auto-surveillance* is the interiorisation of external control, then it would follow that this latter gaze would not even fall under its domain. However, that would make sense only if this latter gaze or encounter, as Nancy refers to it, could come into being. It is at risk of coming into being because it assumes a breakdown of discontinuous selves as Bataille suggests in his description of the erotic imperative. (see Ch.3 n9) All of the tension elicited in counter-acting one's own indefiniteness that Nancy and Bataille's descriptions point to, as well as the *resisting nude's* dilemma described in this study require the present, 'the thing cut by the present which distances it from itself-by the plane that colors and exhibits it.' It is in this fragile present, that the discontinuity is broken down. As the distancing occurs, something gets dissolved. As observed with the *deixis* of the kiss, conditions that permit exceeding identities, break down discontinuity and spill over into other possibilities imply a presence.

Bataille stressed that the transition from the normal state to that of erotic desire presupposes a partial dissolution of the person as s/he exists in the realm of discontinuity. Continuity is what we are after.[11] This continuity is achieved by exceeding being through the indefiniteness of the *present Absence of the Other* as characterized by *the resisting nude*. Keeping in mind that the *resisting nude* battles against its own indefiniteness as it moves out toward something not itself, we are reminded that the 'something not itself' may or may not be the Other. The more important point is that a discontinuous being is not capable of this movement; it is trapped within its own separateness. Separateness is what allows for the ease with which surveillance can be carried out, in discontinuity, the lines are heavily drawn and it is simple to discern one from another. In continuity, discontinuity is dissolved and the person no longer enjoys the alleged safety of separateness. In the case of *auto-surveillance* it becomes more complex since there is an insistence on discontinuity while the Other either remains irrelevant, or rather treats oneself as the Other; the end result is that continuity is still not possible, thereby reducing chances for the erotic.

Nancy, in his elegant poetic vision, and Heidegger before him[12] insisted that individuals are in no way fundamentally separated from each other. "Our existence as singular plural beings is something Nancy repeats like a mantra throughout his work, both tacitly and explicitly. One's relations to others is as primordial to one's existence as one's relation to oneself. ."[13] This philosophical revelation does offer us a defense against the post-modern conceptions of the fragmentation of the self, and it is clearly true, but in light of the kind of surveillance currently in operation, the singular plural being reads like the *resisting nude,* an aberration. Ironically, the singular plural being is now a more radical entity than first conceived by Nancy. It seems that the feeling of profound continuity of being that Bataille stresses as a substitute for individual isolated discontinuity present in eroticism[14] may only be possible as a form of resistance in the twenty-first century.

The sheik stood up, selected the plumpest grape from the best bunch and lifted it to Lamia's mouth. She parted her lips.

Amin Maalouf

CHAPTER SIX

Resistance, Revolt and the Poetry of Jouissance

The medicalisation of everyday life that is a result of Armstrong's Dispensary translates in contemporary terms as a normalization, that is, a stabilizing feature in a global arena where sameness is increasingly valued. Abnormality and deviation in Armstrong's Dispensary could easily be picked up and acted upon. This is true now as we move around under surveillance from the outside in. Coupled with the finding that continuity of being is hardly achievable in this environment while self-possession or the maintenance of stable individuality is a safe, healthy armor against dispossession, the obscene quality of the erotic comes to surface as a resistance. The obscene, as Bataille noted, upsets the physical state associated with self-possession.[1] Dispossession, or the opposite of self-possession is a form of transgression that ultimately destabilizes, upsets the order of things and causes conflict simply through its being.

Julia Kristeva observes that stability is provisional and that in conflict, there is pleasure. "The individual, in his return to him or herself, experiences division, conflict, pleasure and jouissance in this fragmentation."[2] This corporal and cerebral disruption she describes is an indicator of freedom, the freedom to disengage from the stable. As the potential for disrupting stability becomes more and more suppressed, conflict evaporates and takes with it a climate conducive to interrogation and revolt. She follows this line of thinking to the following observation about the technocratic society:

> It is precisely a technocratic ideology that is supposed to abolish anxiety. But what I am saying is the opposite: anxiety, repulsion, nothingness are essential aspects of freedom. That's what revolt is. When one abolishes revolt that is linked to anxiety and rejection, there is no change. You store things and keep storing. It's a banker's idea, not an idea of a rebel, which spreads the technocratic ideology.[3]

The obscene, a stimulator of dispossession and a destabilizer signals anxiety, transgression and the kind of change which is unbearable in this technocratic society. Kristeva's observations lead to the conclusion that while the chances for revolt look bleak in the twenty-first century, there may be a way through by creating a "liberated form of representation of revolt" which would imply a new cultural space open to anxiety, interrogation and conflict.[4] She used the example of young people revolting in the French suburbs and the subsequent problem of encroaching religious dogmatism vs. spiritual anxiety to make a case for creating a cultural space that would allow a mind set on revolt to become a "permanent voice on a level of esthetics."

If, as this study leads to, eroticism may only exist as a form of resistance in contemporary culture, especially under the permeating global surveillance in operation, then the space for revolt becomes increasingly significant. Kristeva's idea of an esthetic of revolt may be paralleled to the aesthetic of eroticism; if the erotic comes to surface as resistance, then in the absence of this revolt of body and mind, there is simply little hope for the erotic. Her esthetic of revolt is based on the possibility of a new cultural space that allows for conflict and destabilization. Since the technocratic society works in opposition to these goals, this new cultural space would have to exist outside of it, or within it, as a subversive micro-society.

Kristeva's "liberating form of representation of revolt" is dependent on the opening of a new cultural space and while this may be possible for the more public anxiety concerning the immigrant population in French suburbs, it seems less likely with eroticism, even though Bataille speaks at length about the correspondence between the private domain of eroticism and religion.[5] The difference here is that in Kristeva's example, the religious dogmatism is intermingled with the current heated political issue of immigration in France and extends the conversation beyond what Bataille linked, that is, the inner experience binding the religious and the erotic. What further problematizes the creation of a Kristevian "new cultural space" in the case of eroticism is the obvious effect of heightened *auto-surveillance* that we saw in the previous chapter. Furthermore, it is increasingly difficult to find this space, in the contemporary domain of a global digital technocratic society.

Kristeva makes reference to Hannah Arendt's idea of anxious thinking [la pensée inquiète] that is in direct opposition to calculated thinking which is simply "computerizing."[6] The storage of data, the repetition of images, the patterned, coded forms of titillation comply with the computerization of thought that Arendt argues against. All of the tension exhibited by the *resisting nude* fighting its own indefiniteness, the disruption by desire, that is, the disruption, the destabilizing that desire unleashes as it moves toward continuity of being, the movement toward the impossible, the unidentifiable pleasure of the kiss, the untouchable of touch can only register as resistance in this landscape. The global digital technocratic society seems like a landscape of freedom as noted in Chapter 4, but it is not a terrain conducive to revolt. It is a 'safe' surveilled place and the kind of societal transgression inherent to eroticism obviously counteracts safety. Transgression is transitory, a moving dynamic which obliterates a fixed space.

Therefore, ultimately in the case of eroticism, we are looking for a space which simply continues to reform itself and therefore remains elusive. It shares with Kristeva's esthetic the allowance for conflict and destabilization. However, it does not have a fixed quality that could fit into a culturally designated frame. Its space is the one that is created at the moment and beyond the moment, the space that allows making an access be as we saw in Chapter 3 in what Nancy calls the 'immemorial pluperfect.' All of it is a proposition, a bareness in motion, dissolving the kind of definiteness and safety that can be guaranteed in a static, controlled environment and as such, is performatively, a resistance.

This brings us back to the two kinds of resistance found at play in eroticism. There is the external one, dealing with the socio-political dimensions of the *resisting nude* itself in the world (Chapter 1) and the internal resistance it experiences as it battles against its own indefiniteness. As we have seen, this act of self-de-limiting creates the resistance the nude embodies. The erotic, then, is a cerebral and corporal revolt fueled by these resistances. So by suggesting that the chances for revolt become more and more precarious in contemporary society, then what or where is the outlet for the erotic, the culmination of resistance?

This study leads to the conclusion that it can only exist through some subversive force. It would have to be something that transgresses 'reasonable' desire or the coded, performance of pornography for example. The only possibility is *jouissance*. *Jouissance,* in its fullest sense is the ultimate transgression that remains impossibly untouchable. However, the Anglo lexicon can't even name it and in some cultures, as in America, it is completely untranslatable. In the glossary of *Sexual Subversions,* there is the most comprehensive explanation of the elusive quality of *jouissance*:

> This term tends to remain untranslated in English texts because of its ambiguity in French. The term refers undecidably to pleasure understood in orgasmic terms, and a more generally corporeal, non-genital pleasure. Sometimes translated by 'bliss', the term does not, however, carry the religious associations of the English term. Sheridan provides the following description. '"Enjoyment" conveys the sense, contained in *jouissance,* of enjoyment of rights, of property etc. Unfortunately, in modern English, the word has lost the sexual connotations it still retains in French. "Pleasure", on the other hand, is pre-empted by "plaisir" . . ."Pleasure" obeys the law of homeostasis that Freud evokes in "Beyond the Pleasure Principle", whereby, through discharge, the psyche seeks the lowest possible level of tension. "*Jouissance*" transgresses this law and, in that respect, it is *beyond* the pleasure principle.'(Sheridan, in Lacan, 1977a,x).[7]

In defining *jouissance* as something that surpasses the Freudian pleasure principle, it also goes beyond the principle of constancy[8], that is, the tendency to keep intracerebral excitation constant. This first appears in Freud's 'Project' under the name of 'neuronic inertia' and he wrote that the principle of constancy "actually was inferred from the facts which forced us to adopt the pleasure principle."[9] The tendency toward constancy lowers tension in its movement toward pleasure and stability. *Jouissance* instead moves toward the highest level of tension found in enjoyment, enjoyment found in pleasure, a corporal, cerebral, social pleasure that has no boundaries and that works its way to both ends of the spectrum, Eros and Thanatos. It is *beyond* "Beyond the Pleasure Principle" even though Freud did counter balance the pleasure principle with the reality principle which actually allowed for a postponed pleasure of satisfaction in order to tolerate feelings of pain, and eventually moved toward recognition of the dynamics of the death instinct. However the death instinct led him back to a return to the earliest state, a desire for total inactivity, 'neuronic inactivity.' *Jouissance* transgresses inertia; it is always turbulent even when still. As such, it is the poetic armor of the erotic.

In speaking about the importance of conflict to an atmosphere ripe for revolt, Kristeva observes that pleasure **and** *jouissance* are instrumental in destabilization. This distinction and inclusion is important; pleasure seeks the lowest possible level of tension and *jouissance* moves beyond any homeostatic limit. This explanation also covers the sexual enjoyment that goes beyond pleasure and as soon as this boundary evaporates, time and space open up, so the provocative transgression Bataille describes that does not deny the taboo but transcends it and completes it or Nancy's poesis-making that remains "extremely elsewhere" and is not brought back alive[10] become viable. *Jouissance* allows for excess; it is the embodiment of desired transgression. It is ironic and revealing that this subversive force is not identifiable in English. The fact that there is no translation may indicate that this level of destabilization and dispossession is not recognizable in some cultures. Of course, that raises the question, does it even exist?

This query brings us back to the destabilized, conflict-ridden *resisting nude*. Bataille remarks that nakedness is dispossession and over and again in this study, the internal and external resistance of the *resisting nude* became a necessary condition for eroticism. In *The Sexual Life of Catherine M.*, the narrator describes the total posing and dispossession that characterizes nudity permitting a heightened level of receptivity. "My nudity feels more complete to me out in the open than in a closed room. When the surrounding temperature, whatever it may be, can be felt by an area of skin it doesn't normally reach, such as the small of the back, the body no longer presents an obstacle to the air, it is penetrated by it and is, therefore, more receptive."[11] This bareness of the *resisting nude* implies a resistance to its own indefiniteness as we saw in Chapter 2. When the bareness is suppressed, there is no chance for resistance. The simultaneity of being completely bared/dénudé and denuded/l'être-à-nu, that bareness in motion that dissolves any definiteness and secures only dispossession and instability can never happen.

It is now plausible to suggest that the *resisting nude* and the transgression of *jouissance* are necessary to the erotic. In landscapes where one or neither of these exists or where there is significant suppression of these for the overall effect of well-being, stability and security, the erotic can not breathe. Perhaps as Millet writes about the physical sensation of the air on the nude and the nude on the air and where Derrida reads Nancy, "touching presupposes a penetration" physically and mentally, and "at the same instant when I am invisibly touched by the other, without any possible reappropriation, which is what I earlier termed absolute mourning"[12] there is a connection which might explain what works against continuity, dispossession and instability and in favor of all those agents that result in a conflict-free landscape. Has the exposure to air, this penetration from the air, all the touch that nudity exacts externally and internally, this ecstatic mourning, is it a threat in contemporary Western culture, especially in the puritanically-rooted United States of America? It has always been, and now under heightened surveillance, as we have seen, it casts a far more threatening shadow.

In the post-Freudian aporias of mourning, Derrida's link between touch and mourning signals the Bataillian observation we heard earlier in Chapter 3, that eroticism is an assenting to life up to the point of death. If touch can be aligned with mourning, then death has already occurred. However, the fear may be more complicated than a simple movement away from the death instinct that is obviously connected to the erotic. "Individual death is but one aspect of the proliferative excess of being."[13] It is the excess of being, the red-flag of jouissance, the modus operandi of the *resisting nude*, the boundary-breaker that could never fit into the 'good patient/ responsible citizen' Dispensary model. At one point, where Bataille recounts some of Sade's violent desires, he writes about the documented appalling cries that accompanied his orgasms and admits that: "I do not know if we are justified in associating these outbursts with mere pleasure. At a certain point, excess can no longer be gauged."[14] Granted, Sade is an extreme example; however, Bataille's point does have some resonance here; how does one gauge excess of any kind, most of all, excess of being? Therefore, a fear of excess of being is a fear of the limitless, a fear of impossible possibilities and works against continuity and dispossession in order to secure a conflict-free landscape. On a basic level, in some countries, such as in the United States, this fear is so elevated that the bareness of the whole *resisting nude* must be obliterated from sight. Fearing the penetration by air, the invisible touch and the instability of excess of being; the erotic has been ushered out leaving its responsible citizens relishing their stability and safety in a post-mourning state.

I'll make you twist them with pain and pleasure until I take you panting, mouth to mouth, and if you don't surrender, I'll impale you . . .

Guillaume Apollinaire

Conclusion

As a Franco-American thinker, I contemplated most of the ideas in this book with a dual lens writing most of it with the Bastille in full view. I mention this as an after- thought to the theory of eroticism I have tried to construct throughout this study, yet everything I have said here pays homage to this influence.

The revolution of *jouissance* described in the last chapter necessitates, firsthand, its recognition and secondly, subsequent human danger and socio-political vulnerability. This could all begin from the benign, fundamental disorder of the caress which is already threatening, as Derrida reads Nancy, "Touching, then, is a question of life and death."[1] Derrida reaches back to Aristotle's discourse on the living and poses the following:

> . . .animals also die when an excessive intensity of touch touches them . . . Couldn't one say that this *measure*, this moderation of touch, remains at the service of life to the sole *extent*, precisely, that some kind of reserve holds it on the brink of exaggeration? A certain tact, a "thou shalt not touch too much," "thou shalt not let yourself be touched too much," or even "thou shalt not touch yourself too much," would thus be inscribed a priori, like a first commandment, the law of originary prohibition, in the destiny of tactile experience.[2]

The reserve that holds touch on the brink of exaggeration as Derrida observes, this law of ordinary prohibition is the tactile's equivalent to the excess of being which must also be restrained in a conflict-free landscape. Even on the most basic, superficial level, we can see this at play in how physical touch gets translated depending on cultural norms. For example, Americans have all kinds of prohibitions and rules of conduct about touch while Europeans and Latin Americans often find themselves holding back from a touch on the shoulder, on the hand, or a kiss on either side of the face when in the company of Americans.

The excessive intensity of touch is dangerous and as we have seen with the *resisting nude*, carrier of all that bareness into the im/possible present; it leads to a destabilization, a disruption and conflict. Yet, as Kristeva points out, conflict is necessary for revolt. Quelling the potential conflict keeps us safe from revolt too. Without the transgression embodied in *jouissance*, without this cerebral and corporal revolt, eroticism disappears. So, the landscape may be safe and stabilized, but everything is contained and excess, "that factor that sets being beyond the limits of definition"[3] is gone.

As soon as a landscape becomes impossible for revolt, it becomes less free and can no longer sustain the erotic.

The United States currently finds itself at this point in the 21st Century. When Kristeva talks about the '68 revolution, she says that "what was at stake was achieving this impossibility-the reality of *jouissance*" and the only way a constraining society would go in this direction would be "if we subject it to fervent and sustained disruption."[4] This happens in France, even today because *jouissance* is a matter of both private and public domain and remains functional. "I revolt, therefore we are . . . still to come."[5] Unfortunately, where the 'I' and 'we' remain discontinuous, the revolt and the erotic never surface. Freedom, in any operative sense, slowly fades out of sight.

Afterword

An alert reader will find the subtext in this study alarming and it should be. When cerebral, corporal and socio-political revolt become impossible, there is a stasis and deadening that spills into all dimensions of life in a free society. When we can't bear the sight of the whole nude body or the touch of the stranger, attempt to 'normalize' all the currents of pleasure and pain, block transgression through homage to order, control and security, mistake Web-Porn as a site of subversion, continue to marginalize all *unreasonable* desire, practice *auto-surveillance* as a daily mantra and over-medicate to stamp out anxiety and conflict, we move toward a deadening comfort zone that distills the erotic and its revolt.

Perhaps it seems counter-productive to work through a

theory of eroticism to ultimately declare its apparent extinction. I would like to think of this study as an aesthetic *brave new world*, a wake-up call to the zones of our beings that have been infiltrated, the ones that separate us from sheep.

I end here with Huxley's question, what can be done? His response in his frightening speculative fiction, faced with threats to humanity and psychological engineering, is to resist. I too arrive at *resistance*, continual conflict-driven *resistance* as the only possibility for sustaining the erotic. Yet, in the case of eroticism, *resistance* and *jouissance* must 'self-touch' [*se*-toucher] à la Nancy and that remains an unnerving, ecstatic pairing.

Notes

Chapter One

1. Gilles Deleuze, *Negotiations,* (New York: Columbia University Press, 1995), p. 54.
2. Susan Bordo, *Unbearable Weight: Feminism, Western Culture, and the Body,* (California: University of California Press, 1993).
3. Judith Butler, *Gender Trouble: Feminism and the Subversion of Identity,* (New York: Routledge, Chapman and Hall, Inc., 1990), p. 25.
4. Michel Foucault, *The History of Sexuality. An Introduction. Vol.1,* (New York: Vintage Books, 1990), p. 67.
5. T. Armstrong, "Dispensary" in *A General Account of the Dispensary for Infant Poor,* (London: Gryphon Editions, 1995), p. 79.
6. Ibid.

7–8. Elizabeth D. Harvey, ed. *Sensible Flesh: On Touch in Early Mod-*

ern Culture. (Pennsylvania: University of Pennsylvania Press, 2002).

9. T. Armstrong, "Dispensary" in *A General Account of the Dispensary for Infant Poor*, (London: Gryphon Editions, 1995), p. 81.
10. Matt Saha, Interview with Director Bernardo Bertolucci, Bullz-Eye, 14 February 2004, 6 May 2007, *<http://www.bullzeye.com/mguide/interviews/2004/bernardo_bertolucci>*
11. George Bataille, *Eroticism*, trans., Mary Dalwood, (London: Marion Boyars Publishers, Ltd., 2006), p.17.
12. Susan Bordo, *The Male Body: A New Look at Men in Public and Private*, (New York: Farrar, Straus and Giroux, 1999), p. 173.
13. Ibid., p.174
14. Ibid., p.253–4
15. Ibid., p.179.
16. Michel Foucault, "The Eye of Power" in *Power and Knowledge: Selected Interviews and Other Writings 1972-77*, ed. Colin Gordon, (New York: Pantheon, 1980), p.148.
17. Susan Bordo, *The Male Body: A New Look at Men in Public and Private*, (New York: Farrar, Straus and Giroux, 1999), p.225.
18. This is a quote from Jim McBride who runs the site that chronicles nudity in film, <http.://www.mrskin.com>.
19. Susan Bordo, *The Male Body: A New Look at Men in Public and Private*, (New York: Farrar, Straus and Giroux, 1999), p.175.
20. Michel Foucault, "The Eye of Power" in *Power and Knowledge: Selected Interviews and Other Writings* 1972–7 77, ed. Colin Gordon, (New York: Pantheon, 1980), p.151.

Chapter Two

1. George Bataille, *Story of an Eye*, (San Francisco: City Lights Books, 1987), p. 40.
2. Ibid.
3. Jean-Luc Nancy, *Multiple Arts: The Muses II*, ed., Simon Sparks, (California: Stanford University Press, 2006), p. 92.
4. Ibid.

5. Ibid., p.106.
6. Ibid., p.105.
7. Jacques Derrida, *On Touching: Jean-Luc Nancy,* trans., Christine Irizarry, (California: Stanford University Press, 2005).
8. Jean-Luc Nancy, Corpus, Editionsié, (Paris: Métaillé Sciences Humaines, 2002). Derrida uses *Corpus* throughout his reading of Nancy's *toucher.*
9. Jacques Derrida, *On Touching: Jean-Luc Nancy,* trans., Christine Irizarry, (California: Stanford University Press, 2005), p. 290.
10. George Bataille, *Eroticism,* trans. Mary Dalwood, (London: Marion Boyars Publishers, Ltd., 2006), p.17.
11. Jacques Derrida, *On Touching: Jean-Luc Nancy,* trans., Christine Irizarry, (California: Stanford University Press, 2005), p.78.
12. Levinas, Emmanuel, Basic Philosophical Writings (Studies in Continebtal Thought), eds., Adriaan Peperzak, Simon Critchley and Robert Bernasconi, (Indiana: Indiana University Press, 1996), p.89.
13. Ibid.
14. George Bataille, *Eroticism,* trans., Mary Dalwood, (London: Marion Boyars Publishers, Ltd., 2006), p. 19.
15. Ibid., p.17–18.
16. Jacques Derrida, *On Touching: Jean-Luc Nancy,* trans., Christine Irizarry, (California: Stanford University Press, 2005), 75.
17. James Waddell, *Erotic Perception: Philosophical Portraits,* (Maryland: University Press of America, Inc., 1997), p.72.
18. Jacques Derrida, *On Touching: Jean-Luc Nancy,* trans., Christine Irizarry (California: Stanford University Press, 2005), p.83.
19. Ibid., p.290.
20. Ibid., p.307.
21. Ibid., p.306.
22. James Waddell, *Erotic Perception: Philosophical Portraits,* (Maryland: University Press of America, Inc., 1997), p. 86.
23. Jacques Derrida, *On Touching: Jean-Luc Nancy,* trans., Christine Irizarry (California: Stanford University Press, 2005), p. 305.

24. Jean-Luc Nancy, *Corpus*, Editionsié, (Paris: Métaillé Sciences Humaines, 2002), p.36.
25. Jacques Derrida, *On Touching: Jean-Luc Nancy*, trans., Christine Irizarry (California: Stanford University Press, 2005), p.274.
26. Ibid., p.306.
27. Ibid., p.307.
28. Ibid.
29. Ibid.
30. Jean-Luc Nancy, *Multiple Arts: The Muses II*, ed., Simon Sparks, (California: Stanford University Press, 2006), p.8.
31. George Bataille, *L'expérience intérieure*, (Paris: Editions Gallimard, 1954), p.131.

Chapter Three

1. Jean-Luc Nancy, *Nous avons voué vie a des signes*, ed. Jean-Paul Michel (Bordeaux: William Blak, 1996).
2. George Bataille, *Eroticism*, trans., Mary Dalwood, (London: Marion Boyars Publishers, Ltd., 2006), p.11.
3. George Bataille, *Literature and Evil*, trans., Alastair Hamilton (London: Marion Boyars Publishers, Ltd., 2001), p.16.
4. George Bataille, *Eroticism*, trans., Mary Dalwood, (London: Marion Boyars Publishers, Ltd., 2006), p.9.
5. Ibid.
6. Jean-Luc Nancy, *Multiple Arts: The Muses II*, ed., Simon Sparks, (California: Stanford University Press, 2006), p.92.
7. George Bataille, *Literature and Evil*, trans., Alastair Hamilton (London: Marion Boyars Publishers, Ltd., 2001), p 44.
8. Ibid.
9. Georges Bataille, *Eroticism*, trans., Mary Dalwood, (London: Marion Boyars Publishers, Ltd., 2006), p. 24.
10. Ibid., p.24–5.
11. George Bataille, *Story of the Eye*, (San Francisco: City Lights Books, 1987), p.33.
12. George Bataille, *Literature and Evil*, trans., Alastair Hamilton

(London: Marion Boyars Publishers, Ltd., 2001), p.43.
13. George Bataille, *Story of an Eye,* (San Francisco: City Lights Books, 1987), p.33.
14. Ibid.
15. Jean-Luc Nancy, *Multiple Arts: The Muses II,* ed., Simon Sparks, (California: Stanford University Press, 2006), p.8.
16. Ibid.
17. Ibid., p.9.
18. Ibid., p.191.
19. George Bataille, L'expérience intérieure, (Paris: Editions Gallimard, 1954), p.26.
20. Georges Bataille, *Eroticism,* trans., Mary Dalwood, (London: Marion Boyars Publishers, Ltd., 2006), p.29.
21. Jean-Francois Fourny, "Writing After the Erotic," *L'Esprit Créateur* (Fall 2004), vol. XLIV, n 3, p.59.

Chapter Four

1. Mario Vargas Llosa, *The Notebooks of Don Rigoberto,* ed., Edith Grossman, (New York: Penguin Books, 1997), p.198–99.
2. George Bataille, *Eroticism,* trans., Mary Dalwood, (London: Marion Boyars Publishers, Ltd., 2006), p.63.
3. Yakini B. Kemp, "Writing Power: Identity Complexities and the Exotic Erotic in Audre Lorde's Writing" in *Studies in Literary Imagination,* (Digital Document, Thomson Gale, Fall 2004), p.2–3.
4. Audre Lorde., *Zami: A New Spelling of My Name,* (California: The Crossing Press, Inc., 1982), p.57.
5. Yakini B. Kemp, "Writing Power: Identity Complexities and the Exotic Erotic in Audre Lorde's Writing" in *Studies in Literary Imagination,* (Digital Document, Thomson Gale, Fall 2004), p.3.
6. Roland Barthes, "The Metaphor of the Eye," *Story of the Eye,* (London: Penguin Books, 2001), p. 127.
7. Ibid.
8. Susan Sontag, "The Pornographic Imagination" in *Styles of*

Radical Will, (New York: Farrar, Straus and Giroux, 1969), p.44.
9. George Bataille, *Eroticism*, trans., Mary Dalwood, (London: Marion Boyars Publishers, Ltd., 2006), p.39.
10. Susan Sontag, "The Pornographic Imagination" in *Styles of Radical Will*, (New York: Farrar, Straus and Giroux, 1969), p.72.
11. George Bataille, *Story of an Eye*, (San Francisco: City Lights Books, 1987), p.64.
12. Susan Sontag, "The Pornographic Imagination" in *Styles of Radical Will*, (New York: Farrar, Straus and Giroux, 1969), p.37.
13. George Bataille, *Eroticism*, trans., Mary Dalwood, London: Marion Boyars Publishers, Ltd., 2006), p.54–55.
14. Ibid.
15. Susan Sontag, "The Pornographic Imagination" in *Styles of Radical Will*, (New York: Farrar, Straus and Giroux, 1969), p.51.
16. Ibid.
17. George Bataille, *Eroticism*, trans., Mary Dalwood, (London: Marion Boyars Publishers, Ltd., 2006), p.16.
18. Michel Foucault, "Preface to Transgression" in *Aesthetics: Essential Works of Foucault 1954–1984*, Vol.2, ed., James D. Faubion, (London: Penguin, 1998), p.28.
19. Jürgen Habermas, *The Philosophical Discourse on Modernity Twelve Lectures*, trans., Frederick G. Lawrence, (Massachusetts: MIT Press, 1990), p.123.
20. Richard Goldstein, "The Double Standard" in *The Advocate*, (Digital Document: Goliath, 20 August 2002), p.1–2.
21. Jean-Luc Nancy, *Multiple Arts: The Muses II*, ed., Simon Sparks, (California: Stanford University Press, 2006), p.21.

Chapter Five

1. Michel Foucault, "The Eye of Power" in *Power and Knowledge: Selected Interviews and Other Writings, 1972- 77*, ed. Colin Gor-

don, (New York: Pantheon, 1980), p. 30.

2. Ibid., p.34.
3. T. Armstrong, "Dispensary" in *A General Account of the Dispensary for Infant Poor*, (London: Gryphon Editions, 1995), p.16.
4. Michel Foucault, "The Eye of Power" in *Power and Knowledge: Selected Interviews and Other Writings, 1972- 77*, ed. Colin Gordon, (New York: Pantheon, 1980), p. 39.
5. Ibid.,74.
6. T. Armstrong, "Dispensary" in *A General Account of the Dispensary for Infant Poor*, (Gryphon Editions, 1995), p.5.
7. Michel Foucault, "The Eye of Power" in *Power and Knowledge: Selected Interviews and Other Writings*, 1972- 77, ed. Colin Gordon, (New York: Pantheon, 1980), p.46.
8. Ibid.
9. Georges Bataille, *Eroticism*, trans. Mary Dalwood, (London: Marion Boyars Publishers, Ltd., 2006), p.18.
10. Jean-Luc Nancy, "Mad Derrida" in *Adieu Derrida*, ed., Costas Douzinas (London: The Birkbeck Institute for Humanities Lectures: Palgrave Macmillan, 2007), p.157.
11. Georges Bataille, *Eroticism*, trans. Mary Dalwood, (London: Marion Boyars Publishers, Ltd., 2006), p.17–8.
12. Martin Heidegger, *Being and Time*: *A Translation of Sein und Zeit*, trans., Joan Stambaugh, (New York: State University of New York Press), p. 12.
13. This is a reference to Nancy's main idea in *Being Singular Plural* which is reflected in his other work mentioned here as well.
14. Georges Bataille, *Eroticism*, trans., Mary Dalwood, (London: Marion Boyars Publishers, Ltd., 2006), p. 15.

Chapter Six

1. George Bataille, *Eroticism*, trans., Mary Dalwood, (London: Marion Boyars Publishers, Ltd., 2006), p.219.
2. Julia Kristeva, *Revolt, She Said: An Interview by Philippe Petit*, trans., Brian O'Keeffe, (California: Semiotext(e) Foreign

Agents Series, 2002), p.100.

3. Ibid., p.101–2.
4. Ibid., p.106.
5. George Bataille, *Eroticism*, trans., Mary Dalwood, (London: Marion Boyars Publishers, Ltd., 2006), p.35.
6. Julia Kristeva, *Revolt, She Said: An Interview by Philippe Petit*, trans., Brian O'Keeffe, (California: Semiotext(e) Foreign Agents Series, 2002), p.114.
7. Elizabeth Grosz, *Sexual Subversions: Three French Feminists*, (Sydney: Allen and Unwin, 1989), p.xix.
8. Sigmund Freud, *Beyond the Pleasure Principle, The Standard Edition*, (New York: W.W. Norton & Co., Inc., 1990), p.6.
9. Ibid.
10. Jean-Luc Nancy, *Multiple Arts: The Muses II*, ed., Simon Sparks, (California: Stanford University Press), p.108.
11. Catherine Millet, *The Sexual Life of Catherine M.*, trans., Adriana Hunter, (London: Corgi Books Transworld Publishers, 2002), p.101.
12. Jacques Derrida, *On Touching: Jean-Luc Nancy*, trans., Christine Irizarry, (California: Stanford University Press, 2005), p. 305.
13. George Bataille, *Literature and Evil*, trans., Alastair Hamilton, (London: Marion Boyars Publishers, Ltd., 2001), p.16.
14. Ibid., p.119.

Conclusion

1. Jacques Derrida, *On Touching: Jean-Luc Nancy*, trans. Christine Irizarry (California: Stanford University, 2005), p.47.
2. Ibid.
3. Georges Bataille, *Eroticism*, trans. Mary Dalwood (London: Marion Boyars Publishers, Ltd., 2006), p.268.
4. Julia Kristeva. *Revolt, She Said: An Interview by Philippe Petit*, trans. Brian O'Keefe (California: Semiotext(e) Foreign Agents Series, 2002), p.36.
5. Ibid., p.44.

Bibliography

Armstrong, T. "Dispensary" in *A General Account of the Dispensary for Infant Poor*. London: Gryphon Editions, 1995.

Arnaud, Claude. "Du Plaisir à l'hypersexe" in *Le magazine littéraire*, n° 470, December 2007.

Bagley, Stephen, ed. *Sex: The Erotic Review*. London: Cassel and Co., 2001.

Barthes, Roland, "The Metaphor of the Eye." *Story of the Eye*. London: Penguin Books, 2001.

Bataille, George. *Blue of Noon*. trans. Harry Mathews. London: Marion Boyars Publishers, Ltd., 2006.

———. *Eroticism*. trans. Mary Dalwood. London: Marion Boyars Publishers, Ltd., 2006.

———. *L'expérience intérieure*. Paris: Editions Gallimard, 1954.

———. *Literature and Evil*. trans. Alastair Hamilton. London: Marion Boyars Publishers, Ltd., 2001.

———. *Story of the Eye*. San Francisco: City Lights Books, 1987.

Blanchot, Maurice. *The Infinite Conversation.* in *Theory and History of Literature.* trans. Susan Hanson Vol. 82. Minnesota: University of Minnesota Press, 1992.

Blau, Marthe. *Submission.* London: Corgi Books: Transworld Publishers, 2004.

Bordo, Susan. *The Male Body: A New Look at Men in Public and Private.* New York: Farrar, Straus and Giroux, 1999.

———. *Unbearable Weight: Feminism, Western Culture, and the Body.* California: University of California Press, 1993.

Butler, Judith. *Gender Trouble: Feminism and the Subversion of Identity.* New York: Routledge, Chapman and Hall, Inc., 1990.

Cars, Laurence and Dominique de Font-Réaulx. *Gustave Courbet: Album de L'Exposition: L'oeuvre de Courbet et sa critique en image.* Paris: Réunion des musees nationaux, 2007.

Deleuze, Gilles. *Negotiations.* New York: Columbia University Press, 1995.

Derrida, Jacques and Elizabeth Roudinesco. *For What Tomorrow . . . A Dialogue.* trans. Jeff Fort. California: Stanford University Press, 2004.

Derrida, Jacques. *On Touching: Jean-Luc Nancy.* trans. Christine Irizarry. California: Stanford University Press, 2005.

———. *The Work of Mourning.* ed. Brault and Naas. Chicago: University of Chicago Press, 2001.

Foucault, Michel. *Discipline and Punishment: The Birth of the Prison.* New York: Vintage, 1995.

———. "Preface to Transgression" in *Aesthetics: Essential Works of Foucault 1954–1984,* Vol.2, ed. James D. Faubion. London: Penguin, 1998.

———. "The Eye of Power" in *Power and Knowledge: Selected Interviews and Other Writings 1972–77,* ed. Colin Gordon. New York: Pantheon, 1980.

———. *The History of Sexuality. An Introduction.* Vol. New York: Vintage books, 1990.

Frappier-Mazur, Lucienne. "Marginal Canons: Rewriting the Erotic." Yale French Studies 75:112–128, 1988.

Freud, Sigmund. *Beyond the Pleasure Principle, The Standard Edi-*

tion. New York: W.W. Norton & Co., Inc., 1990.
Fourny, Jean-Francois. "Writing After the Erotic." *L'Esprit Créateur* (Fall 2004), vol.XLIV, n3, 57–70.
Gallop, Jane. *The Daughter's Seduction: Feminism and Psychoanalysis*. New York: Cornell University Press, 1982.
Goffette, Guy. *Elle, par bonheur, et toujours nue*. Paris: Editions Gallimard, 1998.
Goldstein, Richard. "The Double Standard" in *The Advocate*. Digital Document: Goliath, 20 August 2002.
Grosz, Elizabeth. *Sexual Subversions: Three French Feminists*. Sydney: Allen and Unwin, 1989.
Habermas, Jürgen. *The Philosophical Discourse on Modernity Twelve Lectures*. trans. Frederick G. Lawrence. Massachusetts: MIT Press, 1990.
Harvey, Elizabeth D., ed. *Sensible Flesh: On Touch in Early Modern Culture*. Pennsylvania: University of Pennsylvania Press, 2002.
Heidegger, Martin. *Being and Time. A Translation of Sein und Zeit*. trans. Joan Stambaugh. New York: State University of New York Press, 1996.
Irigaray, Luce. "The Fecundity of the Caress" in *Face to Face with Levinas*, ed. Richard A. Cohen. Albany: State University of New York Press, 1986.
Kemp, Yakini B. "Writing Power: Identity Complexities and the Exotic Erotic in Audre Lorde's Writing" in *Studies in Literary Imagination*. Digital Document: Thomson Gale, Fall 2004.
Kristeva, Julia. *Revolt, She Said: An Interview by Philippe Petit*, trans. Brian O' Keeffe. California: Semiotext(e) Foreign Agents Series, 2002.
———. *Strangers to Ourselves*. trans. Leon S. Rondiez. New York: Columbia University Press, 1991.
Levinas, Emmanuel. *Basic Philosophical Writings (Studies in Continental Thought)*. eds. Adriaan Peperzak, Simon Critchley and Robert Bernasconi. Indiana: Indiana University Press, 1996.
Llosa, Mario Vargas. *The Notebooks of Don Rigoberto*, ed. Edith Grossman. New York: Penguin Books, 1997.

Lorde, Audre. "The Uses of the Erotic: The Erotic as Power" in *Sister Outsider: Essays and Speeches by Audre Lorde*. CA: The Crossing Press, Inc. 1984.

Lorde, Audre. *Zami: A New Spelling of My Name*. California: The Crossing Press, Inc. 1982.

Millet, Catherine. *The Sexual Life of Catherine M.* trans. Adriana Hunter. London: Corgi Books Transworld Publishers, 2002.

Nancy, Jean-Luc. *Being Singular Plural*. California: Stanford University Press, 2000.

———. *Corpus*. Editionsié. Paris: Métaillé Sciences Humaines, 2002.

———. *L'<il y a> du rapport sexuel*. Paris: Editions Galilee, 2001.

———. "Mad Derrida" in *Adieu Derrida*, ed. Costas Douzinas. The Birkbeck Institute for Humanities Lectures: Palgrave Macmillan, 2007.

———. *Multiple Arts: The Muses II*. ed. Simon Sparks. California: Stanford University Press, 2006.

Noys, Benjamin. *Modern European Thinkers: George Bataille: A Critical Introduction*. London: Pluto Press, 2000.

Punday, Daniel. "Foucault's Body Tropes" in *New Literary History*. Vol. 31:3. Maryland: John Hopkins University Press, Summer 2000. 509–528.

Réage, Pauline. *Story of O*. trans. Sabine d'Estree. New York: Grove Press, 1965. (A final suppressed chapter published under title "Return to the Chateau")

Sade, Marquis de. *The 120 Days of Sodomy and Other Writings*. London: Grove Press, 1994.

Saha, Matt. "Interview with Director Bernardo Bertolucci." *Bullz-Eye* 14 February 2004. 6 May 2007 <http://www.bullz-eye.com/mguide/interviews/2004/ bernardo_bertolucci.html>.

Santi, Sylvain. *George Bataille, à l'extremité fuyante de la Poesie*. Paris: Editions Rodopi BV. 2007.

Sontag, Susan. "The Pornographic Imagination" in *Styles of Radical Will*. New York: Farrar, Straus and Giroux, 1969.

Waddell, James. *Erotic Perception: Philosophical Portraits*. Maryland: University Press of America, Inc., 1997.